T0331202

Decoding Black Swans and Other Historic Risk Events

The field of risk science continues to learn from the long history of events to develop principles and practices that enable individuals, organizations, and societies to understand and manage future risk. Reflecting on these histories reminds us that risk and uncertainty are prevalent, yet it remains important to consider what is on the horizon: the possibility of future events, the consequences of those events, our vulnerability to those events, and how to recover from those events.

Decoding Black Swans and Other Historic Risk Events offers a guide to understanding risk events and how to act before they occur. This book explores past risk events and analyzes how risk science principles apply to those events and studies whether current risk science concepts and approaches could potentially have avoided, reduced the impact, or supported recovery following the risk event. New insights are obtained by applying recent research progress in understanding and managing risk, considering aspects including quality of evidence, information, and misinformation in risk studies. The analysis results are used to identify how risk science approaches contribute to the overall management of risk and societal safety, and where improvements can be obtained, allowing the reader to possess a toolkit for identifying and planning for unsafe events.

This title will be a critical read for professionals in the fields of occupational health and safety, risk management, civil engineering, mechanical engineering, energy, marine engineering, environmental engineering, business and management, and healthcare.

Decoding Black Swans and Other Historic Risk Events

Themes of Progress and Opportunity for Risk Science

Shital Thekdi
Terje Aven

CRC Press
Taylor & Francis Group
Boca Raton London New York

CRC Press is an imprint of the
Taylor & Francis Group, an **informa** business

Designed cover image: nancekievill/shutterstock

First edition published 2025
by CRC Press
2385 NW Executive Center Drive, Suite 320, Boca Raton FL 33431

and by CRC Press
4 Park Square, Milton Park, Abingdon, Oxon, OX14 4RN

CRC Press is an imprint of Taylor & Francis Group, LLC

© 2025 Shital Thekdi and Terje Aven

Library of Congress Cataloging-in-Publication Data
Names: Thekdi, Shital, author. | Aven, Terje, author.
Title: Decoding black swans and other historic risk events : themes of
progress and opportunity for risk science / Shital Thekdi, Terje Aven.
Description: First edition. | Boca Raton, FL : CRC Press, 2025. | Includes
bibliographical references and index. |
Identifiers: LCCN 2024019642 (print) | LCCN 2024019643 (ebook) |
ISBN 9781032567631 (hardback) | ISBN 9781032558844 (paperback) |
ISBN 9781003437031 (ebook)
Subjects: LCSH: Risk management.
Classification: LCC HD61 .T47 2025 (print) | LCC HD61 (ebook) |
DDC 658.15/5--dc23/eng/20240805
LC record available at https://lccn.loc.gov/2024019642
LC ebook record available at https://lccn.loc.gov/2024019643

ISBN: 978-1-032-56763-1 (hbk)
ISBN: 978-1-032-55884-4 (pbk)
ISBN: 978-1-003-43703-1 (ebk)

DOI: 10.1201/9781003437031

Typeset in Times
by SPi Technologies India Pvt Ltd (Straive)

Contents

Foreword viii
About the Authors xii
Acknowledgments xiii

Chapter 1   Framework Used to Decode the Risk Events 1

      Works Cited 6
      Further Reading 6

Chapter 2   Historical Events to be Decoded 7

      Global Pandemics 7
         COVID-19 7
         HIV/AIDS 8
         Main Features among Cases 8
      Industrial Accidents 9
         Deepwater Horizon Oil Disaster 9
         Upper Big Branch Mine Disaster 10
         Boeing 737 Max Crisis 10
         Dhaka Garment Factory Fire 10
         Seveso Disaster 11
         Main Features among Cases 11
      Infrastructure Failures 11
         Flint Water Crisis 11
         East Palestine Train Derailment 12
         Main Features among Cases 12
      Natural Disasters 13
         Fukushima Daiichi Nuclear Accident 13
         Texas Energy Grid Failure 13
         California Climate Crisis 14
         Main Features among Cases 14
      Acts of Terrorism and War 14
         September 11, 2001 (9/11) 14
         Main Features 15
      Food Safety 16
         Forever Chemicals 16
         Food Additives 16
         Main Features among Cases 16
         Works Cited 17

Chapter 3   Historical Perspective on Risk 18

      Works Cited 22
      Further Reading 22

Chapter 4   Historical Precedent for Remembering and Understanding
                Risk Issues...23

Chapter 5   Characterization of Surprise and Unpredictability............................26
                Works Cited...30
                Further Reading...30

Chapter 6   Severity of Consequences in Relation to Uncertainties and
                Knowledge..31
                Titanic Case – Evaluation of Consequences......................................33
                Works Cited...37
                Further Reading...37

Chapter 7   Uncertainty Characterizations...38
                Fukushima Daiichi Case – Evaluation of Probabilistic Risk
                Assessment at Nuclear Plants...43
                Works Cited...45
                Further Reading...45

Chapter 8   Types of Knowledge That Inform Understanding of Risk:
                Decoding Black Swans and Other Historic Risk Events....................46
                9/11 Case – Classification of Knowledge..49
                Works Cited...54
                Further Reading...55

Chapter 9   Credibility of Knowledge..56
                Food Additives Case – Credibility of Knowledge..............................58
                Further Reading...60

Chapter 10   Factors Influencing Understanding and Communication of Risk,
                Uncertainties, and Probabilities..61
                Cuyahoga River Fire – Communication of Risk-Related
                Information...65
                Works Cited...71
                Further Reading...71

Chapter 11   Biases, Misinformation, and Disinformation......................................73
                Biases and Moral Hazards – East Palestine Train Derailment
                and Similar Freight Accidents..76
                COVID-19 Misinformation and Disinformation.................................79

Works Cited 80
Further Reading 81

Chapter 12 Balancing Various Dimensions of a Risk Application 82

BP Oil Spill – Dimensions of Risk for Oil and Gas Operations
and Related Regulation 83
Works Cited 87

Chapter 13 The Weight Given to Resilience 88

Hurricane Katrina – Why New Orleans Was Vulnerable and
How Resilience Emerged 89
Works Cited 92
Further Reading 93

Chapter 14 The Big Picture of Risk Science Surrounding Major Historical
Events 94

Tornadoes – The Big Picture Around Risk Issues 94
Smoking, Social Media, and Processed Foods – Linking Three
Global Risk Issues 98
Works Cited 102
Further Reading 102

Chapter 15 How Risk Science Improves the Ability to Address the Gaps in
the Themes Presented 103

Chapter 16 Who Is the Risk Analyst and What Are the Expectations 106

Works Cited 111
Further Reading 111

Chapter 17 Unresolved Issues in Risk Science Identified in the Presented
Themes 112

Abraham Lincoln Assassination – Hindsight Bias versus Noise
Leading to a Risk Event 113
Works Cited 115

Chapter 18 Conclusions 116

Further Reading 117

Index 118

Foreword

Risk events have indelible marks on history, covering decades, centuries, millennia, and beyond. Consider the dinosaur extinction, wars, pandemics, social uprisings, natural disasters, industrial accidents, economic disruptions, terrorist attacks, assassinations, and many others. Even in personal and professional contexts, we have detailed histories of risk events, containing events and conditions that had significant consequences.

While we have documentation for some of these historical events, many remain marked by gaps in understanding, maintain controversial historical accounts, or are otherwise forgotten. The known history provides a basis for our understanding of our world. That history documents both opportunities and negative events that inform our narratives around major problems, controversies, political disagreements, investments, and other initiatives. We all leverage that history when making future projections, decisions, and long-range plans, and when giving advice and communicating with others in our personal and professional lives.

A famous quote of unknown origin states that it can be beneficial to have "good health and a bad memory," suggesting that an over-focus on history can distract us from elements of value today. While that is true from a mindfulness perspective, we need history to inform us as we consider future opportunities and negative events that can impact us in significant ways.

The field of risk science aims to learn from these histories to develop principles and practices that enable individuals, organizations, and societies to understand and manage future risk. Risk science provides a framework to consider what is on the horizon: the possibility of future events, the consequences of those events, our vulnerability to them, and how to recover from them. More specifically, risk science involves principles and practices to characterize risk given available knowledge, support decision-making for risk management, develop initiatives to promote resilience, and aid in risk communication.

With every new risk event, we as a society are faced with two critical questions.

First: *Were risk science principles and practices adequately implemented?*

Second: *Are the current risk science principles and practices sufficient for addressing these types of risk? If not, what are the opportunities to improve capabilities for understanding and managing risk?*

Answering both of those questions is vital. Particularly as the global community continues to develop methods, policies, and practices for adequately handling risk, balancing development, on the one hand, and the protection of values (in particular, related to health and the environment), on the other.

This book aims to address those two questions by applying risk science principles to historical risk events. We then study whether current risk science concepts and approaches could have avoided, reduced the impact, or supported recovery following the risk event. Through the discussion, we identify insights into understanding and managing risk, considering aspects including quality of evidence, information, and misinformation in risk studies. The discussion further illustrates how risk science contributes to the management of risk and safety for societies.

While it is common to think about history as a recollection of events, facts, and data, history is often considered a product of *interpretation*. Our recollection of history is subject to many widely studied systemic factors, biases, and misinterpretations. Some aspects of history get recorded, and others do not. As we attempt to leverage that history for risk science, we recognize that a poor or ineffective understanding of history can potentially result in poor or ineffective risk management. While that is a problematic issue, this book explores how one can practice good risk science even with a spotty understanding of history.

At heart, many of us are already risk scientists. Have you ever:

- Looked at crash test data before purchasing a new car?
- Looked at crime rates in a neighborhood before purchasing a new house?
- Looked at effectiveness and side effects before starting a new medication?
- Considered debates surrounding data, climate change projections, and policy?
- Disagreed with others about the cause or impact of negative events (e.g., accidents and failures)?
- Disagreed with others about the policies proposed by politicians?

All of those tasks, and many others, implicitly involve looking at and interpreting past data and information. There is then some projection of the future, considering the context of the situation. It often becomes difficult to separate those tasks from values and judgment, which can vary from person to person. This book will explore the roles of data, information, values, and judgment in our understanding of risk science so that we can make our own risk-based decisions more clearly and intentionally.

While this is not a history book, we highly leverage historical risk events to decide how principles of risk science relate to those events. Given the magnitude of the consequences of these risk events, they are characterized by feelings of regret. However, we then learn from those events, identifying the main lessons to be learned by exploring the gaps in risk science that may have been a factor in the events' occurrence. We focus the discussion on considering risk science practices and principles to help us all uphold the integrity of risk studies we perform at personal and professional levels.

We use several lenses of risk science to understand those historical risk events. Those lenses are considered throughout the chapters:

Chapter 1 will provide a framework used to decode risk events. This framework is based on recent literature that considers basic principles of risk science and evaluates how those principles relate to past risk events.

Chapter 2 will describe historical risk events to be decoded within each theme discussed later. We consider the following categories of historical risk events:

- Global pandemics
- Industrial accidents
- Infrastructure failures
- Natural disasters

- Acts of terrorism and war
- Food safety

Chapter 3 will provide a historical perspective on how risk has been viewed and treated over time.

Chapter 4 will explore our understanding of past risk events, how we learn from those events, and how our understanding of those risk events can influence our collective risk management now and in the future.

Chapter 5 will explore the role of surprises and unpredictability in understanding and managing risk.

Chapter 6 will explore how we make future projections to understand the severity of a risk event if it were to happen.

Chapter 7 will discuss characterizing uncertainty, how we express uncertainties, the role of knowledge in our understanding of uncertainty, and how we interpret basic characterizations of uncertainties, such as with probabilities and qualitative expressions.

Chapter 8 will discuss the balance between general knowledge and specific knowledge, which helps us understand how the type of knowledge informs the integrity of knowledge that informs a risk study.

Chapter 9 will discuss the credibility of knowledge, including how to identify credible knowledge and promote credibility in both our sharing and interpretation of communications.

Chapter 10 will explore factors that influence the understanding and communication of risk.

Chapter 11 will explore concepts of biases, misinformation, and disinformation that can influence how we interpret and understand risk topics.

Chapter 12 will consider how to balance both the costs and benefits involved with making risk-based decisions.

Chapter 13 will discuss additional practices that enable effective risk management, including robustness and resilience.

Chapter 14 will discuss the big picture for risk science surrounding major historical events.

Chapter 15 will discuss the main gaps identified using the themes discussed earlier. We then explore how risk science improves the ability to address those identified gaps.

Chapter 16 will discuss the role of the risk analyst and potential repercussions for that individual or group of individuals.

Chapter 17 discusses unresolved issues in risk science based on the identified gaps. These issues present opportunities for future research and practice.

Chapter 18 provides an overview of the main issues and gaps explored in this book. This chapter also discusses how one can use the lessons learned and best practices in this book to manage risk more effectively within personal and professional contexts.

While many history books narrate facts and details about historic risk events, this is a book about risk-science-based interpretations of history. We relate historical events to concepts of risk science, including characterization/understanding of risk, probabilistic concepts, black swans, risk perception, and risk communication. No existing work studies those past risk events through the lens of risk science. Interpreting history with the risk science lens translates into stronger principles and methods that can eventually lead to a better understanding/characterization of risk and a better ability to manage risk across applications.

Similarly, while many books discuss principles and methods for risk science, this book is about a risk-science-based interpretation of history. This book does not intend to be a technical book about risk methods. We focus on the practical application of risk science without being overly philosophical or quantitative in our approach. Instead, we aim to bridge the gap between research and application for risk science.

This book will be of interest to undergraduate students, graduate students, industry professionals, and researchers. These audiences will explore how concepts and principles apply to actual risk events and issues. Additionally, audiences will translate practical risk issues into applied and philosophical risk science research.

About the Authors

Shital Thekdi is a professor of analytics and operations at the University of Richmond. She has co-authored many papers on risk management and decision-making. She has a Ph.D. in systems and information engineering from the University of Virginia and an M.S.E. and a B.S.E. in industrial and operations engineering from the University of Michigan. She has several years of experience working in industry, with extensive supply chain management and operational analytics experience.

Terje Aven has been a professor in risk science at the University of Stavanger, Norway, since 1992. Previously, he was also Professor (adjunct) in risk analysis at the University of Oslo and the Norwegian University of Science and Technology. He has many years of experience as a risk analyst in industry and as a consultant. He is the author of many books and papers in the field, covering both fundamental issues and practical risk analysis methods. He has led several large research programs in the risk area, with strong international participation. He has developed many master programs in the field and has lectured on many courses in risk analysis and risk management. Aven is the Editor-in-Chief of the *Journal of Risk and Reliability*, and Area Editor of risk analysis in policy. He has served as president of the International Society for Risk Analysis (SRA) and chairman of the European Safety and Reliability Association (ESRA) (2014–2018).

Acknowledgments

This work builds upon the works of many others in the risk science field. Please refer to the Works Cited and Further Reading sections at the end of this book for references to related work.

AI-based tools were used for fact-checking, sentence structure, grammar, and general proofreading of this manuscript.

We also thank Marisa Crowley for her very helpful feedback and edits during the proofreading of this book.

1 Framework Used to Decode the Risk Events

All risk events are different. This book will demonstrate that risk events occur across geographies, time periods, and societies. They impact widely varying stakeholders and environments. Risk for these events is understood, managed, and communicated using many different practices and contexts, as examined in this book. Some events relate to risk contexts that have been widely studied and understood from a scientific perspective. Some events involve strict legal and professional standards for understanding and managing risk. While the risk science discipline has spent decades developing a general understanding of risk that can be used to characterize and manage risk using a common lens, history may not reflect that common lens. This chapter will closely look at the main components of common understanding across a wide variety of risk contexts.

When considering risk related to some future event, it seems logical to look to the past. We can use our collective history to understand the landscape of future events and their characteristics. This history allows us to study the fundamental questions of risk assessment, as informed by the works of Stanley Kaplan and John Garrick and modified by the authors of this book:

1. What can go wrong?
2. If it does happen, what are the consequences?
3. How likely is it that these events and consequences occur?
4. What is the knowledge and the strength of the knowledge (SoK) supporting these judgments?

We can use these questions to formalize the core ideas of conceptualizing and characterizing risk. Formally, we can consider risk as (A, C, U), where A refers to the events occurring (what is going wrong), C their consequences, and U the uncertainty associated with A and C (will A occur, when, what will C be?). In the risk assessment, analysts characterize risk by (A', C', Q, K), where A' refers to some specified events (aiming at covering the unknown A), C' some specified consequences (aiming at covering the unknown C), Q a way of measuring or describing uncertainties, and K the knowledge that these judgments are based. Typically, Q captures likelihood/probability assessments (P) with associated judgments of the SoK supporting these assessments. Hence, the risk characterization captures (A', C', P, SoK, K), in line with questions 1 through 4 above.

Using this new notation, we can further investigate those fundamental risk assessment questions related to surprise risk events and our collective understanding of history in relation to those risk events. The key point is to understand the difference

DOI: 10.1201/9781003437031-1

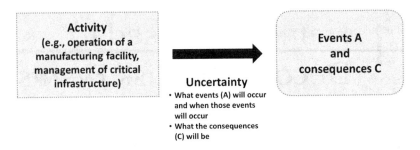

FIGURE 1.1   Illustration of risk concept.

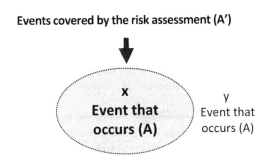

FIGURE 1.2   Illustration of the difference between A types of events and A′. In case the left event occurs (x), it is covered by the risk assessment and A′, but not if the right event (y) occurs. The circle represents the events A′.

between risk (A, C, U) and the risk assessment characterization (A′, C′, P, SoK, K). We illustrate the risk concept in Figure 1.1.

What can go wrong? The true events A occurring could potentially be excluded or overlooked, i.e., A′ does not cover A. See Figure 1.2 for an illustration of this concept. One common reason is that the analysts lack information and/or knowledge about the activity considered and what could happen. Our collective histories can easily cause us to overlook particular events for several reasons. One of the most cited reasons is the lack of observers to document past risk events. Perhaps a risk event A occurred in pre-human times, in settings that lacked documentation, or in settings that included no human witnesses. Another reason for an event to be overlooked is perceived/judged infeasibility, such that an informed person may deem the event impossible or otherwise not worth further investigation from a risk standpoint.

If it does happen, what are the consequences? Translating risk events into consequences requires understanding the risk issue as well as the relationship between the events and the consequences. Understanding this relationship requires knowledge and careful consideration of epistemic uncertainties. Broad generalizations can inform that knowledge about the system and aspects of data, information, assumptions, modeling, testing, argumentation, etc., as will be discussed further in Chapter 9. Again, we can think of situations where some consequences C occurred but were not captured by the specified ones C' in the risk assessment.

How likely is it that these events and consequences occur? The likelihood/probability expresses epistemic uncertainties. Our abilities to record and understand past risk events can help us express the probabilities. Historical data can, to varying degrees, be helpful in predicting future risk events.

What is the knowledge and the strength of knowledge supporting these judgments? Knowledge can emerge in many forms. The most common sources include testing, argumentation, and empirical evidence. However, that knowledge can have varying degrees of strength. Not all data and testing are created equal. For example, consider the difference between scientific study results that have been replicated and widely accepted versus scientific study results that have been contested. Chapter 9 will further discuss the aspects of knowledge.

Figure 1.3 shows the process for general analysis and science. This general process is loosely based on the scientific method. The lower part of the figure shows a complementary process for risk science, which is a subset of general analysis/science. While actual practices may differ depending on the profession, industry, and regulations/standards, the components of the process serve as a generalizable set of tasks that may or may not have been sufficiently performed or documented in relation to past risk events. Using this generalized process, we can more carefully investigate those past risk events using a pre-determined lens that aligns with a general understanding of analysis/science.

For both general analysis/science and risk science, the process begins with a hypothesis and study design. Within analysis/science, this step may involve making an observation, asking a question, or forming a testable explanation. This hypothesis is formed in such a way that it allows for relevant testing, experimentation, and data collection that can be used to form a conclusion. For risk science, this step generally involves a hypothesis related to the characterization of risk discussed earlier in this chapter. Conversely, this step could involve a characterization of system resilience that evaluates the system's ability to respond and effectively recover following a generic risk event. In other specific applications, this step could involve evaluating system vulnerability to particular risk sources/events, robustness to particular risk sources/events, exposure to risk sources/events, or other risk-related terms discussed in this book.

Next, data and information are collected, noting that the pursuit of high-integrity data and information is a major challenge. Data gathering and cleaning are often time-intensive steps in any general analysis/science and risk science initiative. In a risk science context, ensuring that the available data and information are sufficient and applicable to the risk study is also challenging. More discussion about the integrity of data/information is in Chapter 9.

Next, the analysis component consists of analyzing qualitative and quantitative data and information. The analysis could be as simple as an overall conceptual evaluation and as complicated as using rigorous mathematical methods. Typically, the analysis methods are similar and comparable between general analysis/science and risk science. Professional analyses in risk science domains often include modeling and simulation, considering various conditions and risk scenarios.

1. Hypothesis and study design

• *General analysis/science*
  • Proposed idea or assumption

• *Risk science*
  • Level of risk related to the activity
  • (A',C',Q,K) risk characterization
  • System resilience, vulnerability, etc.

2: Data and information

General analysis/science
  • Data/information collection
  • Evaluation of data/information integrity

• *Risk Science*
  • Data/information collection
  • Evaluation of data/information integrity

4. Management review and judgment

• *General analysis/science*
  • Interpretation of analysis
  • Formation of conclusions based on hypothesis

• *Risk science*
  • Interpretation of risk assessment
  • Policy or stakeholder oversight

3. Analysis

• *General analysis/science*
  • Quantiative and qualitative analysis of data and information

• *Risk science*
  • Subconscious evaluation and qualitative reasoning
  • Quantitative reasonining (modeling/computation)

5. Decisions and communications

• *General analysis/science*
  • Communication of study conclusions
  • Decisions based on study conclusions

• *Risk science*
  • Risk management decisions
  • Risk communication

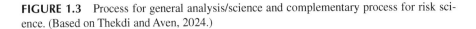

FIGURE 1.3 Process for general analysis/science and complementary process for risk science. (Based on Thekdi and Aven, 2024.)

The management review and judgment stage consists of reviewing the results of the analyses and forming conclusions. From a risk science perspective, this activity includes the characterization of risk and evaluating that risk characterization from various stakeholder or governance viewpoints.

The decisions and communications phase consists of communicating study conclusions and using them to inform any resulting decision-making. When applied to risk science, risk management decisions involve deciding how to address the studied risk. Generally, the options include:

- Accept the risk: This option is chosen if the characterized risk is determined to be sufficiently low.
- Transfer the risk: This option may not always be possible. However, there are cases in which some aspects of the risk can be transferred to other parties, such as through outsourcing and by purchasing insurance.
- Reduce the risk: This option involves investing in initiatives to reduce the risk.
- Avoid the risk: This option involves exiting from some activity that exposes the system to risk.

Risk communication involves sharing risk-related information with various stakeholders, such as the general public, businesses, policymakers, government agencies, and the media. Commonly used methods of communication include reports, advertisements, warning labels, social media campaigns, and public hearings. These communications are not intended to be one-sided. Some risk communication activities, such as public hearings and invitations for comment, seek the perspectives of stakeholders impacted by the risk issue.

Given the characterization of risk within the broader risk management process, the framework of this book explores individual components as follows.

Data and information: Chapter 4 allows us to consider the quality of information in our collective histories as we consider the risk science process. Chapter 5 considers how surprises and black swans factor into our ability to characterize risk appropriately.

Consequences: Chapter 6 considers the difference between actual consequences (C) and the specified consequences (C') during the risk assessment, recognizing that risk science does not necessarily require consideration of every possible consequence, but instead involves understanding the impact of an activity considered and for specific events occurring.

Uncertainty: Chapter 7 contrasts uncertainties expressed using probabilities versus uncertainty judgments using strength of knowledge. These uncertainty judgments may be based on assumptions with a poor basis, leading to the concept of *assumption deviation risk* resulting from incorrect or poorly justified assumptions.

Knowledge: Chapter 8 evaluates knowledge by distinguishing between general knowledge and specific knowledge within a risk characterization. Chapter 9 further expands on the concept of knowledge by considering the credibility of knowledge and evidence in a risk study.

Decision-making and communication: Chapter 11 discusses major emerging issues in the understanding and communication of risk, including issues of bias, misinformation, and disinformation. Chapter 12 explores decision-making as risk science aims to evaluate costs and benefits. Chapter 13 further explores decision-making as activities to address risk, which can also be non-event-specific through investments in robustness and resilience.

By exploring concepts of consequences, uncertainty, knowledge, and decision-making and communication, we can explore how risk history facilitates an improved understanding and handling of risk.

WORKS CITED

Thekdi, S. and Aven, T. (2024). A classification system for characterizing the integrity and quality of evidence in risk studies. *Risk Analysis*, 44(1), 264–280.

FURTHER READING

Aven, T. and Thekdi, S. (2022). *Risk science: An introduction*. New York: Routledge. Chapters 2 and 3.

2 Historical Events to be Decoded

In this chapter, we will discuss the historical events to be decoded in the following chapters. Because we aim to focus on the risk science implications of each type of event, we will focus on the aspects of each event that are relevant, including the features of the event, the consequences, how risk was managed, the available data and information available prior to and during the risk event, and a hindsight characterization of the event.

The description of historical events will follow six main categories:

- Global pandemics
- Industrial accidents
- Infrastructure failures
- Natural disasters
- Acts of terrorism and war
- Food safety

For each category, there are examples of specific risk events. We will discuss the main features of several specific risk events and the generalized features of the collective risk events within each category.

GLOBAL PANDEMICS

Numerous global pandemics have marked human history. Each of them has had tragic consequences. Amidst the catastrophic nature of each pandemic, science, policy, and industry have attempted to make considerable strides in controlling and managing risk, with varying degrees of success.

COVID-19

The most recent and widely discussed pandemic is the COVID-19 pandemic that began in 2019. While the origins of the severe acute respiratory syndrome coronavirus 2 (SARS-CoV-2) are widely debated, it is commonly believed that the virus was first detected in China in late December 2019. Starting in January 2020, many nations enacted travel bans and quarantines for repatriated travelers in an attempt to control spread. Throughout spring 2020, many national public health agencies recommended that citizens undertake measures to protect themselves, communities, and overall populations through mask-wearing, social distancing, and lockdowns. By mid-2020, progress was made in identifying appropriate antiviral drugs and other medical interventions for COVID-19 treatment. By late 2020, vaccines were first

DOI: 10.1201/9781003437031-2

made available to selected populations, while the debate about the virus and the related risk handling, and, in particular, on the effectiveness and safety of these vaccines, continues today. By spring 2023, COVID-19 continued circulating, but the health emergency had ended.

Scientific and academic communities made strides in developing vaccines and medical interventions for pandemic support. Healthcare communities dedicated themselves to treating patients and often put themselves at risk. Taxpayers, industry, and private donors also contributed to funds for vaccine development, public outreach, and pandemic relief.

The consequences of the COVID-19 pandemic were catastrophic, resulting in almost seven million deaths as of the writing of this book. There were many additional tangential effects of the pandemic. Issues included impacts on the global economy, spikes in food insecurity, limited availability of healthcare for medical issues unrelated to COVID-19, insufficient schooling for children due to remote learning, loss of primary or secondary caregivers for children due to virus deaths, issues related to side effects of the vaccines, and many others.

HIV/AIDS

In mid-1981, the first official reporting of acquired immunodeficiency syndrome (AIDS) was published by the U.S. Center for Disease Control (CDC), with the term AIDS used for the first time in mid-1982. By mid-1983, public health agencies developed hotlines and information availability to answer the public's questions and concerns about the disease, reflecting the most modern communication modes available at the time. By the mid-1980s, the cause of AIDS, the HIV virus, was known, allowing for eventual approval for HIV blood testing. During this time, there were many questions about transmissibility and treatment for the disease. Despite being detected in the very early 1980s, it was not until the mid-1980s and early 1990s that the disease was widely discussed, following U.S. President Ronald Reagan's first mention of the disease and following widely publicized cases, including those of children and celebrities like Freddie Mercury and Arthur Ashe. By the mid-1980s, many promising medications were widely researched, with some medications eventually receiving government approval specifically for the treatment of AIDS.

While there appears to be no consensus on total AIDS deaths, the World Health Organization estimates that over thirty-five million people have died from the disease globally, while the virus continues to spread. By early 2024, there was still no approved preventive HIV vaccine.

Main Features Among Cases

While no risk event or context is the same, several underlying factors relate to these pandemics.

First, while these pandemics appeared to be a surprise, especially for citizens with limited knowledge of history and epidemiology, many experts believed that this type of event was possible and plausible. However, entities had varying preparedness for pandemic response and related risk management, which largely contributed to the severe consequences of the pandemics. In Chapter 5, we will further explore these types of surprises.

Second, science and industry had varying degrees of understanding of the origin, features, and treatability of illnesses associated with each pandemic. In addition, some features of illnesses were largely unseen and untracked. For example, it was over a year into the COVID-19 pandemic before widespread testing was available. In other pandemics, there was sufficient relatability to existing knowledge of related diseases to accelerate treatment and management of risk. Chapter 9 will further explore how knowledge relates to these risk events.

Third, risk management for these pandemics relied on citizens. Individual behaviors, such as mask-wearing, vaccinations, and others during the COVID-19 pandemic, sometimes influenced the spread of the illness. Different people interpret risk issues in different ways, whether due to variations in interpretation of science, oversimplification of scientific knowledge, strongly held beliefs, habits, cognitive biases, distrust, or other features to be discussed in later chapters of this book. In addition, population behaviors to address risk rely on a delicate balance between acting with self-interest and the interest of overall populations.

Fourth, these pandemics feature attention on fairness, equity, and equality in the face of a global health crisis. All of the pandemic cases involved illnesses that had non-homogenous impacts across population groups. For example, the COVID-19 pandemic more severely impacted particular demographic groups. Similarly, younger populations were severely impacted by H1N1. These issues of fairness will be further explored in Chapter 17.

Fifth, these pandemics display the importance of multiple stakeholders. The risk management and risk communication efforts involved authorities, policymakers, industry, individual citizens, and other groups. Many individuals with authority remained criticized for spreading incorrect scientific knowledge, particularly in relation to the COVID-19 pandemic. That incorrect knowledge can spread across stakeholders, thereby leading to risk management decisions (including at the policy level) based on poor understanding of the risk issue, which, in some cases, can lead to cascading risk-related consequences.

INDUSTRIAL ACCIDENTS

DEEPWATER HORIZON OIL DISASTER

In the spring of 2010, an explosion in the BP Deepwater Horizon oil rig in the Gulf of Mexico killed 11 workers and injured 17 workers. The environmental consequences were unprecedented, with various sources estimating between three and five million barrels of oil released over 87 days. The spill created a slick encompassing over fifty thousand square miles, later contaminating beaches, marshes, and estuaries. The impact on wildlife and the environment could be detected years after the spill.

The oil spill is attributed to a failed blowout preventer, a critical safety mechanism. Contributing factors included deficiencies in regulatory oversight, cost and time-saving practices, and a deficient safety culture in the organization.

The legal consequences for BP were extensive, with billions of dollars in penalties. These penalties were partly used for environmental restoration. The risk event prompted new regulations and oversight, including new safety measures, requirements for blowout preventers, drilling safety rules, and workplace safety rules.

UPPER BIG BRANCH MINE DISASTER

In the spring of 2010, an explosion in Massey Energy's West Virginia Upper Big Branch coal mine resulted in 29 deaths. The inadequate ventilation system led to the presence of methane, which ignited and caused the explosion. Similar to the Deepwater Horizon case, the risk event is primarily attributed to a poor balance among cost, production, and safety. Many of the alleged causal factors involved violations of health and safety standards.

The CEO faced criminal charges. Later hindsight analyses identified an extensive history of safety violations at the company's mines. The final fallout was also significant, with hundreds of millions of dollars in fines and penalties.

In the aftermath of the disaster, some changes were made in the mining industry. These include improvements in mine inspection procedures, improved oversight, and inspector training.

BOEING 737 MAX CRISIS

In 2018 and 2019, two Boeing 737 Max airplanes crashed, resulting in the deaths of 346 people. Both crashes have been attributed to a sensor failure, with that sensor informing the plane's Maneuvering Characteristics Augmentation System (MCAS). The MCAS, a feature intended to stabilize the plane safely and automatically, contributed to the crashes.

Questions remain regarding how the flawed MCAS system was implemented, particularly whether the inadequate system resulted from basic process-level mistakes in quality testing or more intentional shortcuts. However, workers raised red flags about safety before the risk events. The risk events prompted discussions about whether pilots were adequately trained to override the system, quality practices for new software, and other breakdowns in the process.

Boeing faced billions of dollars in penalties. Since the risk event, after an extensive grounding of the planes, there have been changes to the plane design, including significant changes to the MCAS design. There were also changes in regulations involving pilot training programs and regulations addressing conflicts of interest between manufacturers and regulators.

DHAKA GARMENT FACTORY FIRE

In November 2012, a fire in a garment factory in Dhaka, Bangladesh, resulted in over one hundred deaths. While the cause of the fire is contested, the fire quickly spread through a factory filled with flammable materials. Reports allege workers could not escape the fire due to insufficient emergency exits and escape routes.

While criminal charges were made, the risk event shed light on much more significant issues. This garment factory produced clothing for many well-known companies. The factory had previously been criticized and cited for safety violations, which could ultimately be interpreted as human rights violations within the manufacturing industry. The event also brings to light ethical considerations of brand awareness, ethical sourcing of clothing, working conditions in the garment industry, fast fashion, the pursuit of low-cost clothing, and other consumer habits.

SEVESO DISASTER

In July 1976, an explosion at a chemical manufacturing plant near Milan, Italy, resulted in the release of dioxin. Because dioxin is a human carcinogen, there were severe environmental and public health concerns. After the event, there were reports of health consequences, such as nausea, headaches, and other illnesses. Dioxin poisoning was suspected to be prevalent among residents and wildlife, though data collection on the health consequences is sparse. In the following decades, additional studies shed light on health effects, including chloracne, increased prevalence of cancers, cardiovascular issues, and fertility issues.

The Seveso disaster later became the catalyst for improved safety regulations in manufacturing, particularly in the chemical industry. These directives involved improved accident prevention policies, accident and safety reporting, emergency procedures, and citizen involvement in industrial safety initiatives.

MAIN FEATURES AMONG CASES

First, many of the cases involved the safety culture. Because the safety culture is often difficult to recognize and manage, the issue adds additional complexity to abilities to understand and manage risk. Chapter 10 will further explore factors that contribute to the understanding and communication of risk, which relate to concepts of culture.

Second, many of the risk events involved missed signals that could have prevented or minimized the impact of the risk event. For example, in the Dhaka garment factory fire, the risk event was pre-empted by several safety violations that were left unaddressed.

Third, these risk events involve the balance between efficiencies and risk. In some of these cases, the factors contributing to the risk events directly violated regulation, such as in the Upper Big Branch mine disaster. However, in other cases, such as the Boeing 737 Max crisis, some but not all contributing factors, such as the adequate testing of the MCAS software, involve debate over due diligence in quality testing. Chapter 12 will explore the balance of costs and benefits in the future.

INFRASTRUCTURE FAILURES

FLINT WATER CRISIS

In early 2014, Flint, Michigan, USA, changed its municipal water supply from Lake Huron to the Flint River due to a financial emergency. Following the change, residents complained about the unusual smell and color of the water. Soon after that, a public health crisis emerged due to lead and Legionella bacteria contamination, largely due to the combination of the water supply and water pipes. The lead exposure had lasting and severe impacts on the children of the city, including learning delays, mental health issues, and other health problems. Many citizens also suffered from Legionnaires' disease, leading to multiple deaths.

In 2015, the city changed its water supply back to the Lake Huron source. The fallout for government officials was significant, as the former governor, local officials, and state officials faced criminal charges.

The crisis highlighted water quality and water infrastructure problems across the nation and earth. Several other cities in the United States were also experiencing similar issues with safe drinking water. Meanwhile, across the globe, billions of people experience water scarcity, sanitation, and other issues with water quality and access.

EAST PALESTINE TRAIN DERAILMENT

In February 2023, a train derailment near East Palestine, Ohio, USA, resulted in a massive release of hazardous materials. Several derailed cars contained vinyl chloride, ethyl acrylate, and isobutylene, leading to serious concerns over the derailment's environmental and public health impact.

While no injuries were reported, nearby residents and businesses were evacuated. A few days later, evacuation orders were lifted based on evidence that air and water quality had returned to safe levels. The assertion of safety was widely contested among community members and the media, as the evidence was conflicting, and residents continued to raise health concerns. There were also significant environmental concerns as chemicals were detected in nearby creeks.

In the following weeks, investigations showed a wheel bearing failure contributed to the derailment, noting that the risk event would have been preventable through improved inspection and maintenance. The event prompted allegations of inadequate regulation involving related safety sensors for derailment warnings. In addition, the risk event brought to light other similar derailments that had similar causes but did not attract as much public attention due to their relatively lower ramifications. More broadly, the risk event had other contributing factors, including the many stakeholders and regulators involved with the infrastructure and conflicts between internal safety policies versus broader rail safety rules. When writing this book, the full repercussions were yet to be determined.

MAIN FEATURES AMONG CASES

The case studies highlight several main themes:

First, the causal factors for risk events often balance on a fine line among many system owners, decision-makers, and stakeholders with varying self-interests. In the Flint water crisis example, there appeared to be some conflict of interest between political figures and local officials. In the East Palestine train derailment example, the balance or potential imbalance between the interest of federal regulators and the internal interests of private industry contributed to the disasters. Generally, these imbalances suggest that minimal adherence to rules (regardless of whether rules are interpreted correctly) without considering stakeholders and broader risk implications can contribute significantly to the health, social, and reputational consequences for risk events.

Second, the risk events involved highly regulated infrastructures with large degrees of training, such that the topic areas for risk are well-established. Nevertheless, the risk events occurred. There is recognition that despite these well-established risk processes, risk events can still occur. We see the implications of this issue in the East Palestine train derailment example, given a general sentiment that overall railroad safety had improved in the previous decades due to the advanced technologies and safety procedures. There remain complex questions, such as: "Is a poor risk policy better than no risk policy at all?" or "Are there systemic problems or misalignments in existing risk policies?" or "Do conflicts of interest among various stakeholders reduce the effectiveness of risk policies?" Chapter 15 will further explore how risk science can help address gaps in basic operations to help answer those questions.

NATURAL DISASTERS

FUKUSHIMA DAIICHI NUCLEAR ACCIDENT

In March 2011, a major earthquake resulted in a tsunami, creating waves that damaged the Fukushima Daiichi nuclear power plant operations in Japan. The waves impacted the plant's emergency generators, which are critical for reactor core cooling. Due to nuclear meltdowns, hydrogen explosions, and a radiation release, nearby areas were evacuated. Radioactive water was released into the Pacific Ocean, and evacuated community members continued to be monitored for long-term health issues.

More extensive risk assessment and management procedures may have prevented the disaster. The risk event also brought to light extensive vulnerabilities in infrastructure, as issues like proximity of nuclear reactors to oceans are not easily changed or addressed.

Because the affected area is in a region that is prone to earthquakes and tsunamis, controversy over nuclear power in general is highly politicized. Much attention remains on the appropriate assessment and management of risk for this type of infrastructure.

TEXAS ENERGY GRID FAILURE

In February 2021, severe winter storms disrupted the energy infrastructure of Texas, USA. While the death toll remains contested, some estimate over one hundred deaths, with many deaths due to hypothermia in winter weather conditions combined with a lack of electricity for residents. A public health crisis emerged as the state faced food and water shortages, disruptions in healthcare resources, manufacturing stoppages, and other issues with severe social and economic impacts.

The causes of the infrastructure failure encompass many dimensions. The design and connectivity of the infrastructure itself was highly politicized. The grid failure also came amid the COVID-19 pandemic, when site visits and oversight were virtual and within a context of public-health-related uncertainty. From a more operational lens, the infrastructure failure directly resulted from the failure to appropriately winterize equipment.

The disaster prompted additional legislation to promote the reliability of energy infrastructure during extreme weather events. These practices include improved winterization of equipment with particular attention toward critical facilities.

CALIFORNIA CLIMATE CRISIS

Over the last several decades, California, USA, has faced several severe natural disasters. These natural disasters include rising sea levels, wildfires, floods, and droughts. These risk events have impacted ecosystems, agriculture, and residents. The rising temperatures have contributed to smog that can impact respiratory health. The extreme heat can factor into other health issues like heatstroke and dehydration.

Projections for climate change suggest that severe weather events in California will continue to be pervasive.

MAIN FEATURES AMONG CASES

These risk events highlight several main themes:

First, these risk events involve inherently present vulnerabilities for the entities subject to risk. In the Fukushima Daiichi example, vulnerabilities related to the proximity of nuclear power plants to the ocean are inherent conditions that infrastructure owners cannot control for existing structures. Similarly, in the California climate crisis, the area has minimal control over the climate conditions. Individual entities can do little to reduce the likelihood of a particular risk event. Instead, these entities have some control over how they can manage the risk in acknowledgment of those vulnerabilities. Chapter 13 will further explore the weight given to resilience.

Second, these case studies highlight the need to recognize that risk events can have varying durations. In the California climate crisis, the risk event is presented as a collection of many risk events with related causes and implications. In the Fukushima Daiichi nuclear accident example, the effects of the event are long-term and not concluded at a particular point in time. The full consequences of these risk events are not necessarily known or understood. Chapters 6–8 will further explore the estimation of consequences, uncertainties, and knowledge.

ACTS OF TERRORISM AND WAR

SEPTEMBER 11, 2001 (9/11)

In September 2001, two hijacked planes flew into the World Trade Center in New York City, USA. A third plane flew into the Pentagon in Arlington, Virginia, USA.

FIGURE 2.1 President George W. Bush visits the World Trade Center Disaster Site on September 14, 2001. National Archives (2001) photo no. 5997294.

A fourth plane, alleged to be headed toward Washington, DC, USA, crashed in Shanksville, Pennsylvania, USA.

The attacks resulted in about three thousand deaths. The emergency response and recovery operations at the site of the World Trade Center attacks, labeled Ground Zero, were extensive and dangerous. The heroes contributing to the recovery efforts faced respiratory problems and were prone to cancers as a result of exposures following the risk event.

The disaster affected communities worldwide. In late 2002, the United States Department of Homeland Security was created to address homeland security risk, including terrorism-related risks. Today, across the world, homeland security is operationalized with a large emphasis on the tasks of prevention, mitigation, preparedness, response, and recovery.

Figure 2.1 shows President George W. Bush, other political figures, and emergency personnel touring the World Trade Center Disaster Site on September 14, 2001.

MAIN FEATURES

The 9/11 example illustrates the discrepancies in knowledge. While the terrorist groups had full knowledge of the planned attacks, the U.S. government had some but limited knowledge and warning. This case relates extensively to several chapters, including Chapter 5, which relates to surprise and unpredictability, and Chapter 8, which relates to knowledge.

FOOD SAFETY

FOREVER CHEMICALS

PFAS (per- and poly-fluoroalkyl substances), referred to as "forever chemicals," are dangerous chemicals in industrial and consumer products. Products containing forever chemicals include nonstick cookware, shampoos, cleaning products, food wrappers, and many other commonly used items. As a result, PFAS are widely found in human and animal blood samples. The health effects of PFAS are thought to include problems with cholesterol, fertility, cancers, and other conditions.

While individuals can take actions that reduce exposure to PFAS, there are also initiatives to reduce PFAS from consumer products. However, forever chemicals remain prevalent. It is speculated that these chemicals are not banned due to a variety of reasons, including limited research on the health effects of PFAS, pushback from industry, the necessity of PFAS for products to function, and varying consumer sentiment for the removal of PFAS.

FOOD ADDITIVES

Food additives for coloring, taste, and shelf-life have been used for thousands of years. However, artificial additives have increasingly been used in the last few decades to produce processed foods, personal care products, and pharmaceutical products. These artificial additives have a mixed history of being valued by customers and viewed as a dangerous and unnecessary component of foods.

While there are risk issues with many coloring additives, one specific additive is Red Dye No. 3. This dye was commonly used in cosmetics and foods. The use of the dye has been banned for various purposes in some regions, but not in others. While research on the health effects of this dye is conflicting, there are links to hyperactivity in children and cancers. However, the use of this dye continues as it remains cheaper than natural additives. The dye remains commonly found in food items targeted at children, like sprinkles, ice creams, and candies.

Controversies over the use of food additives often come in waves as new research provides new information on health effects. The new research often conflicts with previous research, further adding confusion. Recent regulations call for bans of some additives and to disclose the use of those additives in products.

MAIN FEATURES AMONG CASES

These risk events highlight several main themes:

First, these cases illustrate issues of risk amplification and attenuation. When new information, new risk events, or increased attention often brings a particular risk issue to public attention, the narrative around the risk issue can change. Later chapters will discuss this phenomenon in more detail. Chapter 10 will further explore related issues with risk communication.

Second, these cases illustrate risk issues in which there is conflicting research and information to support the projection of potential consequences. Later chapters will discuss the main principles, including the precautionary principle, that can guide conversations about risk in these situations. Chapter 9 will further discuss the role of knowledge in understanding risk.

Third, these cases illustrate the role of risk exposure. The health effects of the chemicals in these cases may depend on dosage, propensity for illness, and other factors. However, the health issues are often discussed at a policy level, calling for bans and other blanket-level changes. This raises significant questions about who is responsible for managing risk (e.g., individuals, governments, and others?). Chapter 6 will further explore the severity of consequences, while Chapter 12 will relate risk to the practice of decision-making.

WORKS CITED

National Archives (2001). *911: President George W. Bush Visits New York.* https://catalog. archives.gov/id/5997294

3 Historical Perspective on Risk

The events of this book will be decoded by considering risk principles as they exist today. However, the understanding and treatment of risk has varied considerably over time. In this chapter, we explore how the understanding and management of risk has evolved, particularly in business and society.

This operational and societal perspective is also intertwined with philosophy, religion, and practical considerations. One cannot internalize complex risk topics without leveraging core stances across various aspects, such as belief in a predestined fate, ethical codes of conduct, belief in the impacts of one's actions, and a sense of duty when faced with unpredictable or uncontrollable events. To a large extent, these stances have influenced rules, regulations, and policies regarding risk, as described in this chapter.

At a societal level, ancient civilizations spanning centuries recognized uncertainties and the potential for adverse outcomes. Societies developed roads, bridges, and aqueducts to improve safety and efficiencies in transportation and access to resources. Societies developed security measures by constructing forts, armies, and warfare-related technologies. These investments addressed risk by promoting health, safety, a productive local environment, and access to various resources, particularly for the intended societal stakeholders. While these examples span many centuries of history, and the world has changed and progressed considerably over time, much of this book addresses the same core issues.

As a more specific example, consider ancient agricultural practices developed over the past many centuries. Agriculture was and continues to be a fundamental aspect of society. Agricultural practices were not static but continuously evolving in response to new technological advancements, conditions, experiences, culture, and societal views. Societies faced significant risk related to weather, pests, soil conditions, and consequential crop failures. Interpretations of the causal factors related to these risk sources varied as new understandings and scientific knowledge developed. Through developing experience and knowledge about the phenomena contributing to deficiencies in production, societies documented and studied weather patterns, conditions of plants, and more to help inform assessments of current conditions and to make informed predictions about future conditions. To mitigate risk, societies continuously developed advanced agricultural techniques, such as crop rotation, diversification of crops and livestock, fertilizers, and water management systems like canals and reservoirs. At a broader level, societies developed resilience practices through storage and preservation in preparation for potential food shortages, including drying, canning, smoking, fermentation, and other more advanced processing practices.

DOI: 10.1201/9781003437031-3

Perspectives on operational risk continued to shift during the Industrial Revolution from the late 1700s to the early 1800s. The increased use of machines and large-scale factories for production increased the effort to develop and reconsider relationships between people, machines, and technologies. Factories were known for being unsafe working environments due to their dangerous, complicated machinery. In addition, the factories were often poorly lit, poorly ventilated, and overcrowded. Workers, including children, worked long hours in those unsafe conditions. Science and industry also had poor knowledge of the health impacts of exposure to coal dust exposure, toxic chemicals, and other harmful substances.

In response to the need to more appropriately manage risk related to working conditions, new regulations began to address those concerns. While dangerous working conditions continue to exist across the world today, these regulations present a mindset shift toward using policy to address and manage risk. For example, the nineteenth century Factory Acts legislation in the United Kingdom aimed to improve working conditions and protect worker health and safety. These types of regulations generally aimed to ban the employment of children, improve safety conditions, limit working hours, and improve factory ventilation. In addition, movements promoted the adoption of worker compensation laws, requiring employers to compensate workers and their families after workplace accidents or illnesses.

The risk and safety concerns of the Industrial Revolution also promoted the growth of trade unions and labor movements. These groups often leveraged collective bargaining to negotiate for improved safety initiatives (e.g., equipment, training, staffing, workplace inspection, and monitoring). These movements also called for protections for whistleblowers, potentially encouraging workers to report safety violations and hazardous conditions, thereby further addressing major risk issues.

Worker movements had substantial impacts. For example, the Flint sit-down strike of 1936–1937 in Flint, Michigan, is one of American history's most historically discussed labor actions (see Figure 3.1). Following the Wagner Act of 1935 giving workers the right to form unions, there was an impetus to consider working conditions, job security, and other norms within the manufacturing industry. The General Motors (GM) automobile factory strike resulted from general worker sentiments regarding low wages, long hours, and poor job security. During the strike, autoworkers sat down in factories and refused to work. This was in stark contrast to previous strikes when employees had primarily leveraged picketing. The striking workers effectively shut down production across several GM plants in the area. After the 44-day strike, GM agreed to recognize the United Auto Workers as a bargaining representative for its employees, presenting a significant turning point in the U.S. labor movement.

The Flint sit-down strike and other related movements illustrate the importance of recognizing the perspectives of various stakeholders when understanding and managing risk. The movements also increasingly recognize that risk-based decision-making should not be a top-down process but more collaborative in nature. In other words, important risk decisions and understanding are not solely in the domain of a single few, but instead should be decided in collaboration with those

FIGURE 3.1   Flint sit-down strike 1937. Photographed by Dick Sheldon. Library of Congress (1937) control no. 2017790326.

who are seriously impacted by the risk issue. These issues highlight the importance of considering risk from a more holistic perspective that expands beyond simple financial metrics, instead considering areas of overall safety, quality of life, and ethical concerns.

More formal methodologies to understand and manage risk began to develop in the 1960s. Through collaborations with science and industry, Probabilistic Risk Assessment (PRA) was developed to analyze and quantify risk, particularly within the nuclear and aerospace fields. PRA leverages probability theory and mathematical models to estimate the likelihood and consequences of various scenarios involving system failures, human errors, and risk events. The use of PRA was not without controversy, which will be further discussed in later chapters of this book. The growing use of PRA in the 1970s led to criticism over using relatively new and untested risk methodologies to manage risk for major catastrophic events. While there were specific quantitative reasons for the criticism, the controversy over the use of PRA also stemmed generally from discomfort and skepticism toward quantifying risk for extreme and catastrophic events. This type of risk quantification also required a shift from relying on risk perception versus probabilities and data to making risk-based decisions. This tension between perceived and quantitative risk continues today, as this book will further explore.

Despite the controversy and debate over the use of PRA, its growing use has led to a general movement toward understanding and managing risk using quantitative tools and information at a systematic level across organizations and from a policy perspective. In this new age of risk understanding, PRA contributed to a shift toward integrating evidence-based quantification with subjective judgments for understanding risk. This change led to a more holistic view of risk, allowing for considering multiple factors and scenarios that could lead to risk events, considering both likelihoods and consequences. As a result, there was a culture shift toward understanding and managing risk in transparent and repeatable ways and leveraging that formalized risk approach for more purpose-driven resource allocation for risk management initiatives.

Following the advent of PRA, many other widely used tools and regimes followed. Many were methodological, such as Fault Tree Analysis, Event Tree Analysis, Hazard and Operability Analysis, Failure Mode and Effects Analysis, and many others.

In the 1990s, Enterprise Risk Management (ERM) gained popularity. Using this concept, organizations adopted a more holistic approach combining financial, operational, and strategic risk. The structured frameworks of ERM require organizations to systematically identify risk issues, use structured approaches to assess risk, and use structured approaches combined with stakeholder engagement to prioritize risk management initiatives. Like PRA, ERM promoted a mindset shift toward understanding an organization's risk culture and risk appetite, recognizing that risk management involves values and is not a wholly quantitative exercise.

The September 11, 2001, terrorist attacks had an immense impact on how organizations and societies understand risk. The attacks led to organizations being more aware of the potential for catastrophic terrorist attacks, with particular recognition of geopolitical forces. There was also increasing emphasis on operational continuity in response to various risk scenarios, recognizing the criticality of particular operational functions (e.g., healthcare and emergency response) in the aftermath of disasters. In addition, there was increased attention on critical infrastructures (e.g., energy, transportation, communication, emergency services, and healthcare) and the interdependencies among those critical infrastructures, leading to more intentional understanding and management of risk for these infrastructures that are essential for national security and public safety. The attacks also resulted in increased involvement of policy within risk initiatives. Across the world, new regulations and government agencies promoted a heightened awareness of risk issues, increased accountability, and improved coordination among various stakeholders.

The technological changes of the early 2000s led to increased public knowledge and discourse of complex risk issues. The infamous phrase "If you see something, say something" from the United States Department of Homeland Security post-9/11 appropriately summarizes a public sentiment to be aware and involved in risk activities and risk signals. Those types of initiatives would have been solely the responsibility of specific groups in the past. With that increased public awareness came increased expectations for risk communications to be transparent and credible.

As history has shown, the understanding and management of risk continue to evolve. The COVID-19 pandemic is a reminder that while the public health community was aware of pandemic-related risk issues, the COVID-19 pandemic was largely seen as a surprise by societies in general. The pandemic also exposed risk issues that were widely unaddressed before this time, including the impacts of social media, misinformation, public sentiment, political forces, and interconnectedness on the ability to mitigate population-level risk. Similarly, discourse around climate change is further linking risk to the impacts of slow-onset emergent conditions, activism, social equity, and other issues generally unaddressed decades prior.

While no crystal ball can predict the risk issues on the horizon, there are signals about key unaddressed issues. The evolution of risk understanding has generated a tug-of-war between risk perception and a quantified risk concept. This issue permeates the risk events of this book, as public behaviors and sentiment are largely the product of risk perception. As exemplified during the COVID-19 pandemic, there is momentum toward studying falsehoods that can drastically impact public perception. This suggests that risk perception can be influenced by misinformation, misunderstanding, and politicization related to core knowledge of a risk topic area. There is also increased understanding that particular risks impact populations in different ways. Thus, there is increased discussion about equity (fairness in providing risk-related initiatives by considering individual needs) and equality (all individuals are given the same risk-related initiatives regardless of circumstances) within understanding risk and subsequent decision-making for risk management initiatives.

As the evaluation of risk has shown, risk science continues to grow and develop over time. The following themes presented in this book will further explore this evolution within the context of risk science principles.

WORKS CITED

Library of Congress (1937). Sitdown strikers in the Fisher body plant factory number three. Flint, Michigan digital file from original neg. https://www.loc.gov/resource/fsa.8c28669/

FURTHER READING

Bedford, T. and Cooke, R. (2001). *Probabilistic Risk Analysis*. Cambridge: Cambridge University Press.
Bernstein, P. L. (1996). *Against the Gods*. New York: Wiley.
Covello, V. T. and Mumpower, J. (1985). Risk analysis and risk management: An historical perspective. *Risk Analysis*, 5, 103–120.

4 Historical Precedent for Remembering and Understanding Risk Issues

Individuals and societies extensively strive to understand, remember, and interpret the past, particularly past risk events. Those risk events may be of personal significance, and those events may be extensive, involving entire communities, regions, and the world.

At the individual level, scientists continue to develop an understanding of the human brain's ability to create memories. These memories may change due to time, experiences, individual moods, environments, contexts, disorders, and in response to traumatic risk events.

Businesses and societies can also "remember" the past by considering available data and information collected over time. With the prevalence of video recordings, sensor tracking, and data collected on personal devices, it is clear that risk events can leave footprints that can live perpetually. However, with that past data and information, we collectively remember primarily only what is recorded and how those recordings are perceived and disseminated.

Our interpretations of past risk events can depend on the interpretations from others, such as from news articles, history books, gossip, and social media. Those information sources may have varying motivations for sharing their interpretations and may benefit from some narratives more than others. Some entities may control the narrative around risk events, such as by influencing what society and other specific stakeholders view as causal factors, objects of blame, and impediments to recovery. Thus, those controlling the narrative can influence the attention, energy, and emotional gravity surrounding a risk issue.

Our interpretations of available information and those past risk events can also be misinterpreted or have multiple potential explanations. Consider, for example, a memory of hearing a siren. Individuals may have varying explanations for the siren's origin depending on physical abilities, the explanations of others, limited awareness of surroundings particularly in chaotic situations, information from media sources, the presence of other competing sirens or noise, etc.

Despite these abilities to remember and understand past risk events, we also recognize that our memories and understanding are often incomplete. We individually do not always witness every detail related to the narrative surrounding a risk event. Humans do not generally have a perception of every detail in their sensory frame of reference. Humans also cannot always identify signals, causal factors, or circumstances that could contribute to a risk event. Similarly, the second-hand narratives

DOI: 10.1201/9781003437031-4

recounted by others often omit key details and can contain intentional and unintentional inaccuracies.

As a result of the imperfect information, we tend to fill in gaps in understanding using various tendencies. For example, myths, legends, conspiracy theories, assumptions, and fabrications allow us to make sense of complex events that would otherwise be poorly understood. This phenomenon is not new, as "confabulation" affects many who subconsciously fabricate stories and logical explanations to fill in information gaps.

Regardless of how we remember past risk events, we interpret those risk events through today's lens. We often categorize those memories into good versus bad, right versus wrong, preventable versus non-preventable. Similarly, in risk contexts, we commonly use the term *hindsight bias*, suggesting that when risk events happen, we often overestimate our ability to have foreseen, prevented, or mitigated the risk.

However, hindsight bias is not necessarily wrong. Consider the cases of industrial incidents, including the Upper Big Branch mine disaster and the Dhaka garment factory fire. In many cases, some signals warned about an impending risk issue. For example, repeated documented safety violations signaled a systemic problem. The risk was not appropriately managed if the safety violations were unaddressed or addressed in cursory ways. Similarly, the barriers to whistleblowers warning others about the risk event also undermine any risk management initiatives.

In industrial settings, there is precedent for carefully analyzing past risk events and identifying causal factors. Quality control initiatives, such as performing root-cause analysis and fishbone diagrams are a few of many tools that engineers often use to identify the causal factors for risk events, with the eventual goal of making operational changes that can avoid future risk events. Many professions beyond engineering also undergo similar activities, sometimes called postmortem analyses, intending to improve abilities to understand and manage risk continuously.

Poor risk management does not necessarily result in large-scale risk events. The opposite can also be true – high-quality risk management does not necessarily result in the absence of risk events. In the risk literature, there is significant discussion of near-misses, as the circumstances could potentially lead to a risk event. Due to luck or other external factors, past near-misses did not result in a high-magnitude risk event. While one cannot speculate whether careful analysis of near-misses would have avoided the risk events related to the industrial accidents, understanding the history and circumstances of the near-misses is vital for understanding and managing future risk.

However, it is important to note that those near-misses did not result in a high-magnitude risk event. Humans can also learn from the history of those near misses and incorrectly conclude that the circumstances of the near-miss were not a concern or that the circumstances would likely not end up contributing to a future risk event because no risk event occurred due to the past near-miss. Thus, a misleading sense of security emerges from incorrectly evaluating those near-misses.

Sometimes memories of a particular risk event are somewhat meaningless in understanding future risk events. Consider as an example events that are thought to be memoryless, as the future likelihood (frequentist probability) of a risk event is not related to whether the event has happened in the past. Many common events are thought to exhibit memoryless properties, such as coin flips, computer hardware failures, and arrival rates. Thus, overreliance on quantifying the frequency of some risk events can mislead us into making incorrect conclusions about the likelihood of future events, recognizing also that the likelihood of future events can be consistently changing. Consider, for example, the concept of the 500-year flood. If a 500-year flood occurred today, that does not necessarily mean the next one will not happen until 500 years from now.

There is often an opportunity to learn from risk events and effectively implement risk assessment and management regimes to prevent future events. However, politics, bureaucracies, speed of regulation, and interpersonal factors can impact efforts to do so. Also, those risk issues are not isolated from the roles of business, law, politics, individual freedoms, fairness, and other factors that influence real-life decisions.

The factors discussed in this chapter suggest that our individual and collective memories are unreliably unreliable. Memories may be of high quality and accuracy in some cases, completely false in some cases, and logically correct with key inaccuracies in others. Regardless of the accuracy of the memory, it takes intentionality to understand where accuracies and falsehoods exist in the information we use to understand and manage future risk.

In the following chapters, we will look at those accuracies and inaccuracies and how they relate to abilities to understand and manage risk sufficiently. We will investigate the role of knowledge, the strength of knowledge, and key properties of evidence that contribute to high-quality risk science.

5 Characterization of Surprise and Unpredictability

The future is unwritten – it is unknown to us today. While we can make projections using our sense of intuition, patterns we have recognized over time, data and analysis, and the informed opinions of others, one can never be entirely certain of a particular future outcome.

Similarly, ruling out particular outcomes, even those that seem impossible, is difficult. Surprises can happen. Unforeseen events do occur. Many major risk events of our time, such as the 9/11 terrorist attacks, came as a surprise. That is, these events were a surprise to most of us but not necessarily to all stakeholders involved with the surprise event. We will discuss the aspect of knowledge and the imbalance of knowledge in later chapters.

Surprises and unforeseen events relate to the concept of the "black swan," originating from the poet Juvenal. According to this history, there was an Old World assumption that all swans were white, as only white swans had been witnessed and recorded. However, this knowledge was updated when an expedition led to the discovery of black swans in Western Australia.

Nassim Taleb (2007) further categorizes the black swan concept using three properties:

- The event is an outlier outside the realm of regular expectations.
- The event carries an extreme impact.
- Humans can create explanations for the event's occurrence after the fact, allowing for the event to be explainable and predictable.

More generally, we can say that a black swan is a surprising and extreme event relative to one's knowledge/beliefs. These could be events that are entirely unknown from a scientific perspective, events that were not considered in a risk analysis process, and events that are not believed to occur because the probability of the event was judged to be very low.

We reframe the black swan concept into three different interpretations, as discussed in Aven (2015).

a. An unknown unknown (events unknown to all) with extreme consequences (to be further explored in Chapter 8).
b. An unknown known (events unknown to some but not others) with extreme consequences.

DOI: 10.1201/9781003437031-5

c. A surprising extreme event that is not believed to occur due to a very low probability and related relatively strong knowledge judgments.

Consider, for example, the 9/11 terrorist attacks. These attacks were a black swan because they were a surprise relative to what was believed or known at the time. They can be viewed as both b) and c) types of black swan.

The 9/11 event was conceivable in the sense that there was some intelligence providing signals that an attack was possible. However, government officials did not clearly communicate or appropriately understand those signals. In contrast, the event was not conceived, as officials did not sufficiently address or act on those signals. It could have been conceived if sufficient processes and communication were in place before the event.

Regardless of the complexity of the event, black swans emerge when surprises happen that are relative to one's knowledge/beliefs. In these cases, the existing knowledge could seem rather strong from the perspective of the risk assessor, as in the case of the white swans 300 years ago. However, when considering the knowledge of many other viewpoints, this knowledge was, in fact, poor, such that there was a poor understanding of the system and its behavior.

The black swan concept highlights the important role of the knowledge basis surrounding a risk event. For example, consider two possible risk events, A_1 and A_2. Suppose they were both assigned the same probability of occurrence, which was determined to be 0.00005%. The probability assigned to event A_1 was supported by high-integrity information, including detailed and trusted data, and also supplemented by an understanding of the main causal factors for the event. Suppose the probability assigned to event A_2 was based on low-integrity information, such as heresy from unqualified information sources. As a result, surprises in reference to A_1 may be very different from surprises in reference to A_2, yet both of these events can be considered black swans.

Another metaphor for a rare and surprising event is the perfect storm, defined as a rare confluence of well-known phenomena creating an amplifying interplay leading to an extreme event. There are many theories on the origin of the *perfect storm* as a term, but we will briefly describe one popularized origin related to a meteorological event. In this case, the remnants of a tropical storm merged with a low-pressure system, combined with the effects of a high-pressure system that blocked the storm's movement. The conditions intensified the storm's strength, resulting in extremely high waves and strong winds. While all three weather conditions were well-known and familiar, the combination was rare. However, this particular combination of conditions caused an extreme weather event. Those in a risk assessment role for a fishing boat, the Andrea Gail in 1991, decided that the water conditions were adequately safe, given the low potential severity of each single weather condition, but neglected to foresee the impact of all three weather conditions combined. Consequently, the extreme weather event resulting from the combination of weather conditions led to the sinking of the boat and no survivors.

In risk terms, a perfect storm involves a combination of events A' leading to an extreme event. This perfect storm event may also be a surprise event relative to one's knowledge/beliefs. The individual phenomena contributing to a perfect storm may not have the potential to create a high-magnitude risk event, but these individual events combined have the potential for great harm. These perfect storms often are events that can be predicted with relative accuracy, events that can be somewhat predicted with some accuracy, and events with unpredictable circumstances.

Consider, for example, the Fukushima Daiichi nuclear accident. The event was not unforeseen, as earthquakes and tsunamis were known phenomena. Regulations and knowledge were also considered during the site development, but that existing knowledge declared the impact of a possible tsunami to be low for that particular coastline. However, that knowledge was updated prior to the event, warning that the potential impact of a tsunami was significantly higher than previously thought. While the new knowledge was broadly known, the event seemed unlikely to many, and the site was primarily left unprepared for the event. The concurrence of these events (earthquake, tsunami, and new knowledge) amplified the consequences. While the site had procedures to address each risk separately, it was unprepared to handle the combination of events. Thus, the Fukushima Daiichi nuclear incident can be seen as the result of a perfect storm.

While the COVID-19 pandemic is often referred to as a black swan event, it is sometimes also referred to as a perfect storm event. While the virus and disease itself were a surprise to many, pandemics of large magnitude have been witnessed in the past and pandemics have continuously been studied and discussed within public health and policy domains. As such, the pandemic did not come as a surprise and should not be labeled a black swan. Many believed the likelihood of a pandemic of similar consequence to COVID-19 was considerable. The pandemic itself was also combined with other ongoing risk issues and conditions that exacerbated the event. In particular, there were impacts of the propensity and mechanisms to spread falsehoods, which can be categorized as misinformation and disinformation (further discussed in Chapter 11). In a time of high uncertainties, such as with a new virus, it is expected that information from various sources will be met with distrust or skepticism and be labeled a falsehood. The messages and narratives related to the falsehoods created confusion about the origins of the virus, how the virus spreads, how to treat the disease, and many other aspects of the risk event. However, those issues with misinformation and disinformation were not new and were not generally thought to be major threats to public health prior to the pandemic. This time period was also a time of growing social media use, which was used to spread those falsehoods. The combination of misinformation/disinformation, social media use, and the virus exacerbated the impact and duration of the pandemic risk event. Thus, the pandemic can be viewed as a perfect storm.

Another metaphor used to describe rare, surprising, and extreme events is the *dragon-king*. The term *king*, introduced in this context by Laherrère and Sornette, describes extreme outliers as compared to past witnessed patterns. In risk terms, a dragon-king is an extreme event resulting from emergent behavior caused by the amplifying interplay of a confluence of uncertain factors. A dragon-king can also be viewed as a black swan. Dragon-kings often occur with events that are understood using existing data and information. These events often slowly progress toward instability, sometimes referred to as emergent conditions. Consider again the 9/11 terrorist attack risk event. There was existing knowledge about the formation of terrorist groups enmity against Western society. It was also a time of rapid development of the Internet, which increased the capabilities of those terrorist groups to communicate and recruit. In addition, while government agencies had intelligence staff, much information about hijacking threats was largely unstudied and unaddressed. The interplay among those factors, including emergent conditions (e.g., Internet), contributed to the risk event.

Table 5.1 summarizes the metaphors discussed in this chapter. These concepts are further discussed by Glette-Iversen and Aven (2021). All of these metaphors broadly contain similar elements of the black swan, but we also show nuanced differences among those metaphors.

Given the issues with the surprises, perfect storms, and dragon-kings, it is clear that history alone is not sufficient for understanding and managing future risk. There needs to be more forward-looking thought and analysis related to knowledge about the risk. There is a need for intentionality in defining critical needs for knowledge, understanding what types of knowledge are relevant, and understanding the strength of available knowledge. For example, when managing risk in reference to surprises, we consider when to enact a cautionary/precautionary strategy to take mitigative measures when there are large uncertainties related to knowledge and understanding of the risk. Similarly, there may be a need for a discursive strategy to reduce uncertainties and increase knowledge concerning the risk.

As we discussed in this chapter, while surprises can happen in the form of black swans or perfect storms, these surprises are in relation to one's knowledge and beliefs. Therefore, properly characterizing the knowledge/belief through historical risk in combination with understanding the relevance of historical perspectives is critical in managing risk related to these events. We will discuss this issue further in the following chapters.

TABLE 5.1
Classification of Metaphors Discussed in This Chapter

Metaphor	General Interpretation	Example Event
Black swan	A surprising or extreme event relative to present beliefs/knowledge	9/11 terrorist attacks
Perfect storm	A rare confluence of well-known phenomena creating an amplifying interplay leading to an extreme event (can be viewed as a black swan)	Fukushima Daiichi nuclear accident; COVID-19 pandemic
Dragon-king	An extreme event resulting from emergent behavior caused by the amplifying interplay of a confluence of uncertain factors (can also be viewed as a black swan)	9/11 terrorist attacks

WORKS CITED

Aven, T. (2015). Implications of black swans to the foundations and practice of risk assessment and management. *Reliability Engineering and System Safety*, 134, 83–91.

Glette-Iversen, I. and Aven, T. (2021). On the meaning of and relationship between dragon-kings, black swans and related concepts. *Reliability Engineering and System Safety*, 211, 107625.

Taleb, N. N. (2007). *The Black Swan: The Impact of the Highly Improbable*. London: Penguin.

FURTHER READING

Aven, T. (2014). *Risk, Surprises and Black Swans*. New York: Routledge.

Paté-Cornell, E. (2012). On "black swans" and "perfect storms": Risk analysis and management when statistics are not enough. *Risk Analysis*, 32, 1823–1833.

6 Severity of Consequences in Relation to Uncertainties and Knowledge

In earlier chapters, we discussed conceptualizing and characterizing risk in the risk assessment process. We characterized risk using the notation (A', C', Q, K), where A' was some specified events, C' was the specified consequences, Q was a measurement or description of uncertainties, and K was the knowledge that these judgments are based. The term Q can capture likelihood/probability assessments with associated judgments of the strength of the knowledge (SoK) supporting these assessments, leading to the risk characterization (A', C', P, SoK, K).

In this chapter, we discuss the consequences C'. Those consequences serve as a prediction of the future that is informed by available data, information, assumptions, modeling, argumentation, etc. The consequences are not in isolation to Q and K. In fact, it would be negligent from a risk perspective to specify C' without consideration of Q and K. The text of this chapter will explore C' only, but refer the reader to Chapter 7 to further explore uncertainty characterization and also to Chapter 9 to further explore concepts of knowledge.

These predictions for C' can be wrong and widely contested due to many factors, including uncertainties, knowledge, opinions, deception, and many others. We will further explore the legitimacy of those predicted consequences in this chapter.

First, it is important to clarify how consequences are understood and measured. Metrics that have been used to model consequences include dollars, health, environment, sustainability, and other more specific and detailed metrics. It is also common to create objective functions that combine these types of metrics into a single formula. For example, consider the objective function:

$$\text{Objective} = \text{Profit} + \text{Societal Health} + \text{Environment}$$

In general, there is a broad objective for the system studied. Most decision-making aims to increase or meet a particular goal for a chosen objective. For example, suppose a clothing manufacturer uses the objective function shown here with a wide perspective on profit and other societal and environmental considerations. This company is very focused on societal health in relation to the manufacturing of their project. Particularly, in response to the Dhaka garment factory fire risk event, there is a large concern over worker safety, fair wages, and the elimination of child labor.

DOI: 10.1201/9781003437031-6

Additionally, there is concern over the environmental impact of their manufacturing and of the finished product clothing. There have been broad accusations that others in the industry have polluted rivers and streams with harmful chemicals. The company itself has been accused of being a mechanism for fast fashion, encouraging customers to quickly replace their clothing, which will eventually end up in landfills.

The Profit term in the objective function can be easily quantified in units of dollars.

The Societal Health metric is not so easily quantified because societal health involves conditions often hidden or overlooked by policymakers, such as in the case of the Dhaka garment factory fire. However, the company could quantify this metric by considering the percentage of manufacturing facilities in their supply chain that passed a safety inspection. They may also supplement that information with site visits, talking to employees, and soliciting information from community members.

The Environment metric could be quantified using metrics from an engineering analysis of pollution by the manufacturing facility, using metrics like milligrams per liter for contamination. Alternatively, this metric could incorporate metric tons of greenhouse gas emissions.

The function can also be reframed to contain a single standard metric, such as dollars:

Objective = Profit + Dollar Value of Societal Health + Dollar Value of Environment

Reframing into dollars can result in many uncomfortable conversations about quantifying qualities like health, safety, the well-being of people, the well-being of the environment, and other things we value. However, monetizing these aspects can be convenient for mathematical purposes. While there is controversy over whether monetization is fair, the monetization does support transparency in how estimates are made.

The term *value of a statistical life* (VSL) or *willingness to pay* is used to quantify consequences involving human life. VSL represents the cost individuals would be willing to pay to reduce the number of expected fatalities by one. Similarly, other monetized metrics, like quality-adjusted life year, incorporate quality of life by considering factors like state of health and the amount of time individuals experience a particular state of health. The dollar value of the value of a statistical life differs considerably among government agencies and the private sector. The exact value is often in the range of $5 million to $15 million. The exact language for these terms also tends to differ, with some government agencies choosing terms like "willingness to pay to reduce the risk of dying."

While these VSL-related terms are often accused of putting dollar values on human lives, the tool's intent is to aid in risk reduction. The metric is one factor out of many that can help make decisions about the most appropriate initiatives to address risk. Generally, these metrics intend to evaluate the benefits (in monetary terms) of avoiding fatalities, thereby allowing for the comparison of various risk reduction activities and policies. For example, consider the role of a transportation agency considering various safety improvements in an area (e.g., crosswalks, crossing signals, and sidewalks). The transportation agency can use VSL to prioritize and expedite those safety improvements in areas where they would be most effective.

When using a single standard metric containing multiple dimensions, one must ask whether overperformance in one metric can compensate for underperformance in another. For example, one can ask whether exceeding expectations in environmental health can compensate for underperformance on metrics involving human health. This a value judgment that should always be understood and discussed by decision-makers.

When considering consequences in a risk assessment process, the risk analyst, decision-makers, and other stakeholders are forced to make a judgment on which aspects of risk matter. As expected, choosing a single metric can be troublesome as different stakeholders may view the risk and associated consequences differently from others. For example, consider the Dhaka garment factory fire example. A risk science perspective would carefully weigh the importance of the health and safety of workers in comparison to profit and efficiencies.

Across all of the historic risk events studied in this book, very few took place in settings with a systematic risk assessment and risk management regime that resulted in a prediction of consequences. However, even without a systematic risk assessment and management regime, risk was assessed and managed using informal techniques, such as intuitive decision-making, socially constructed norms, internal policies, and adherence to regulatory requirements. In cases in which little to no regulation existed surrounding the risk issue, there may have been large variability in risk assessment and decision-making regimes.

Forward-looking projections for consequences also heavily rely on assumptions. Current operating conditions, such as current climate conditions or current economic conditions, often inform these assumptions. These assumptions can also be based on what is widely believed about potential risk events as informed by recent histories and memories.

The following Titanic case study more closely studies how regulations and assumptions informed the ship's design.

TITANIC CASE – EVALUATION OF CONSEQUENCES

Consider the case of the infamous Titanic, owned by the British White Star Line. This passenger liner, thought to be unsinkable by many, struck an iceberg in the Atlantic Ocean and sank during its maiden voyage in April 1912. Despite leveraging the latest design principles and technologies of the time period, the Titanic lacked sufficient lifeboats for all passengers and crew. As the ship sank, more than 1,500 out of the about 2,200 passengers onboard lost their lives due to inadequate lifeboat capacity.

Consider regulation, assumptions, and consequence projections involving the number of lifeboats in this case. There were 20 lifeboats that could hold about 1,176 people. This was clearly a problematic design. The vessel's capacity was about 3,300 people, far above the capacity of the available lifeboats. An early design of the Titanic included 64 lifeboats, then reduced to 32, and finally further reduced to 20 (16 lifeboats and 4 collapsible). Figure 6.1 shows passengers on a ship's deck with lifeboats in the background. It is widely believed that the decision to reduce the number of lifeboats was for aesthetic and recreational purposes to service the first-class passengers.

FIGURE 6.1   Ship deck similar to Titanic on April 12, 1912. National Archives (1912) photo no. 278332.

While a formal risk policy did not appear to be in place in the Titanic case, there are signs that the ship's design was compliant with the relevant regulations of the time, when smaller ships were the norm. When the Titanic was built, the design was subject to the Merchant Shipping Act of 1894, which required the number of lifeboats to be determined by considering the ship's tonnage. The largest ships required at least 16 lifeboats, which the Titanic exceeded.

There are several underlying assumptions about potential consequences considered during the design of the Titanic.

First, the Titanic was deemed to be "unsinkable." Despite the prevalence of ship sinkings at the time, it was believed that the 16 compartments of the hull were watertight. The ship could stay afloat even if multiple compartments flooded. Passengers likely focused their attention on the grandeur and luxury of the Titanic and assumed the state-of-the-art ship met, or possibly exceeded, regulatory requirements.

Second, some viewed lifeboats as unnecessary on the presumably unsinkable ship. In a worst-case scenario disaster, lifeboats would be of little use. The designers may have accepted the risk associated with that worst-case scenario disaster. Instead, the available deck space that could be used for additional lifeboats may have been viewed as being more valuable for recreation purposes.

Third, the designers may have focused on limited scenarios in which the lifeboat design would have been sufficient. For example, lifeboats could be used and reused to transport passengers, so they carry passengers to a destination, then return to the ship and pick up additional passengers. In the case of the Titanic, the RMS Carpathia rescued the passengers who were granted access to the lifeboats. However, that rescue took place long after the ship sank. The actual events of the Titanic did not meet the assumptions in the planned scenario and the different type of assumed risk event. Figure 6.2 shows Titanic passengers in lifeboats in near-freezing temperatures.

FIGURE 6.2   Titanic passengers on lifeboats. National Archives (1912) photo no. 278336.

FIGURE 6.3   Senate Investigating Committee questioning J Bruce Ismay of the White Star Line. Library of Congress (1912).

The underlying assumptions about the necessity and use of lifeboats greatly fueled the devastating consequences of the tragic risk event. Figure 6.3 shows the head of the White Star Line being questioned about the decisions leading to the tragic risk event.

Unfortunately, it was not until after the risk event that regulation and practice further improved the requirements for lifeboats on sea vessels. In 1913, the International Conference for Safety of Life at Sea called for lifeboat space for all passengers, recordkeeping for the opening and closing of watertight doors, inspection of those watertight doors, lifeboat drills, radiotelegraph installation, a certificated watcher for receiving and understanding radiotelegraph distress signals, and a multitude of other safety enhancements. In addition, the Safety of Life at Sea Convention agreement in 1914 prompted the International Ice Patrol and subsequent improvements in international cooperation to avert similar disasters.

While the Titanic case study illustrated the need to consider a wider variety of scenarios and consequences during the ship design, the following years of history illustrate that understanding and managing risk with consideration of consequences generally requires learning from history and being vigilant about changing conditions. One case point is the RMS Carpathia, which rescued the Titanic passengers in 1912. A few short years later, in 1918, the ship sank during World War I after being torpedoed by a German submarine. Potential consequences of torpedo attacks were known and somewhat addressed at the time. As a result, there were attempts to manage the risk of these attacks, as the ship traveled with a convoy to avert submarine attacks. Hours after an escort left the convoy, a torpedo was sighted. Despite attempts to avoid attack, the attack by three torpedoes in total led to the deaths of five people.

In contrast to the Titanic example, current formal risk assessment regimes in government and industry model consequences using more data-informed and quantitative methods. For example, consequences can be modeled using simulation and mathematical representations of systems. With those more current methods, just as in the case of the Titanic, there is a need to make assumptions within those methods. Sensitivity analysis and other related mathematical tools, out of the scope of this book, can help understand how the model reacts to changes in assumptions. This is one way of doing a knowledge-check as a sensitivity analysis can help understand how the consequences may differ when key assumptions, such as those based on poor knowledge, deviate.

In addition to modeling assumptions, understanding the consequences for risk applications requires careful consideration of scenarios. As with all future predictions, scenarios involve considering a wide variety of factors that could change. Scenarios could be slow-moving changes in conditions, like those involving climate change and economic cycles. Scenarios such as earthquakes, natural disasters, and pandemics could also be sudden and severe. For example, in the Titanic example, scenarios could involve evacuation, assuming lifeboats could be reused versus not being reused during an emergency.

It is also important to consider the time element of consequence projections. In the Titanic example, the design of lifeboats and assumptions about potentially reusing lifeboats did not consider the potential for the risk event to occur over many hours. Due to the prolonged time lifeboats were at sea, reuse of lifeboats was not a possibility.

Another example considers the Flint water crisis projection of short- and long-term consequences. In the short term, water quality issues led to issues with customer complaints. In the long term, those consequences morphed into serious long-term health repercussions and deaths. Some may see incentives to be shortsighted about considering consequences, as regulatory and reporting requirements can also be in the short term. However, strong risk assessment and risk management regimes must consider both short- and long-term consequences.

Across the examples explored in this chapter, many uncertainties related to the risk topic existed. Those uncertainties translate into a lack of clarity in projecting or estimating both short- and long-term consequences associated with risk events. That uncertainty will be further discussed in Chapter 7.

WORKS CITED

National Archives (1912). Uncaptioned photograph of a ship deck that was similar to the Titanic. https://catalog.archives.gov/id/278332

National Archives (1912). Photograph of two lifeboats carying TITANIC survivors. The following caption appears on the back of the mat: "Boat No 14, 5th Officer Lowe, Mrs Compton, Mys. Compton, Mrs. Minahan, Mys. Minahan, Mrs. Collyer, Mys. Collyer, W.I. Hoyt (who died), about 25_, Towing Engelhardt-D.". https://catalog.archives.gov/id/278336

Library of Congress (1912). TITANIC disaster. Senate Investigating Committee questioning individuals at the Waldorf Astoria. https://www.loc.gov/pictures/item/2002721386/

FURTHER READING

Aven, T. and Thekdi, S. (2022). *Risk Science: An Introduction*. New York: Routledge. Chapters 8 and 9.

7 Uncertainty Characterizations

Uncertainty is an important aspect of risk. This chapter looks into the challenge of characterizing uncertainties. We show how this characterization is linked to probabilities, knowledge, variation, and imprecision.

There are different types of probabilities. The most common are the following:

- **Classical probability**: This type of probability applies when the situation studied involves a finite number of outcomes (say N), and each outcome is equally likely to occur. Then the probability of a specific outcome is $1/N$, while the sum of those probabilities over all outcomes equals 1. Suppose you have a six-sided, fair die, similar to those used for board games or in gambling activities. The outcomes can be 1, 2, 3, 4, 5, or 6 (one for each side of the die). Then, the probability of rolling any specific number will be 1/6. When this type of probability is applicable, there are many opportunities to leverage the assumptions to make elaborate probabilistic models.
- **Frequentist probability**: This type of probability is used when we construct (in real life or as a thought exercise) a large (in theory, infinite) number of similar situations to the one considered. The probability of an event, A, is defined as a fraction representing the time that this particular event occurs in this population of similar situations. One example is in the context of mass production of a bolts to be used for airplane doors. Then, the frequentist probability of a unit being in a failure state (i.e., the bolt does not meet quality standards) is the proportion of bolts in a failure state (i.e., not meeting quality standards).
- **Subjective (knowledge-based, judgmental) probability**: This type of probability is often used in applications when there is little data or the data is not representative of future assumptions. Instead of using past data, this probability captures uncertainty or degree of belief for some event to occur. The probability is based on the available knowledge and is interpreted as: Suppose you assign the probability of an event A, P(A) to be 0.30. This means that the entity assigning the probability has the same uncertainty and degree of belief for event A to occur as randomly selecting a red ball out of an urn containing 100 balls, within which 30 balls are red. Suppose K represents the knowledge that is the basis for the assigned probability. In that case, we write P(A|K) to show the dependency of the probability on K, which we could say as "the probability of A given K." An imprecise subjective probability is an interval, for example, at least 0.30. It is interpreted

DOI: 10.1201/9781003437031-7

similarly, but the urn now has not 30 red balls but at least 30 balls. The imprecise probability then allows for the assigner, who may be hesitant to provide exact probabilities, to be less precise.

The subjective probabilities express epistemic uncertainties, which relate to issues with knowledge of the system and any associated modeling and assumptions. There are many sources of epistemic uncertainties that will eventually be reflected in P, K, and judgments about the strength of knowledge in a risk study. Data is an example. We recognize that the data itself is certain – it is static and known. However, there are issues concerning the volume of the data (e.g., sample sizes, time periods, and geographies), the quality of the data (e.g., sources and trust), and the relevance of the data (i.e., appropriate for the study) in reference to the phenomena studied. The data could also be used with assumptions that contain fundamental errors or may be used in inappropriate or incorrectly executed models. When the data and associated modeling is used for forecasting or any future projection, the phenomena described by the data are only one component of a variety of factors that can influence future events and conditions.

When data comes from expert opinions about uncertainty, those opinions may be conflicting. The experts are experts in their field but may not necessarily be experts about risk and uncertainty. The conflicting data can also result from different views on the matter studied. When there is conflict in the elicitation of expert opinions about uncertainty, this can be reflected in the strength of knowledge judgments.

In addition to the use of the term *epistemic uncertainty*, another frequently used term is *aleatory uncertainty*. However, the reference to the term uncertainty in this context is unfortunate, as the term expresses variation in phenomena. This type of variation does not represent the uncertainty of those assessing the risk. Frequentist probabilities and probability models (e.g., normal distribution) express variation and are the basis for statistical methods, such as confidence intervals, hypothesis testing, machine learning, and artificial intelligence. Consider, for example, a 95% confidence interval for the average concentration of pollutants in a contaminated creek. If we select an infinite number of water samples, 95% of the confidence intervals from those samples would contain the true population average contamination.

The use of probabilities in formal risk assessments has a long history. Beginning back in the 1960s and 1970s, there was momentum for using probabilities for risk assessment in new and groundbreaking technological initiatives like space travel and nuclear power. NASA made early progress in this area by attempting to quantify the probability of sending astronauts to the moon and returning safely as part of the Apollo project.

During the decades of early development of space travel and the growth of nuclear energy, there was continued interest in computing the probability of system failures. Agencies experimented with the use of probabilities in risk assessment through the use of Probabilistic Risk Assessment (PRA), Failure Modes and Effects Analysis (FMEA), and other related tools.

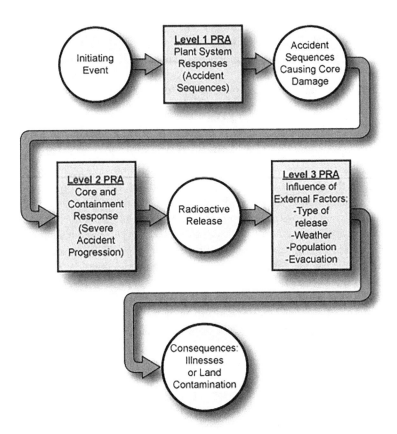

FIGURE 7.1 Probabilistic risk assessment consequence analysis (one component of PRA process) used within nuclear industry. Nuclear Regulatory Commission (2024).

As an example, consider the PRA methods used today. Figure 7.1 shows an example of one component of the PRA process as used by the U.S. Nuclear Regulatory Commission. This consequence analysis step considers a specific initiating event, consideration of specific core damage situations, external factors, and consequences to people and the environment.

The WASH-1400 Nuclear Reactor Safety Study by the Nuclear Regulatory Commission, published in 1975, was an early adopter of the use of PRA for nuclear power plants. At the root of the PRA was the intent to use data and existing knowledge to compute the probability of risk events, such as those related to the core damage. However, at the time, the nuclear industry was relatively new. There was little history of nuclear operations that could be used to substantiate those probabilities.

Only two decades earlier, in 1958, the Shippingport Nuclear Power Station, near Pittsburgh Pennsylvania, became the first commercial nuclear energy facility in the United States. President Dwight D. Eisenhower, who was present for both the groundbreaking and the opening of the power station, was a strong advocate for the nuclear industry. Earlier in 1953, his famous "Atoms for Peace" speech to the United Nations

General Assembly called for using nuclear science to peaceful initiatives and those initiatives with economic potential. In the speech, he stated:

> The United States knows that if the fearful trend of atomic military build-up can be reversed, this greatest of destructive forces can be developed into a great boon, for the benefit of all mankind. The United States knows that peaceful power from atomic energy is no dream of the future. The capability, already proved, is here today. Who can doubt that, if the entire body of the world's scientists and engineers had adequate amounts of fissionable material with which to test and develop their ideas, this capability would rapidly be transformed into universal, efficient and economic usage?
>
> To hasten the day when fear of the atom will begin to disappear from the minds the people and the governments of the East and West, there are certain steps that can be taken now.

The speech alludes to the world's intense memories of nuclear weapons used in World War II. While the authors do not speculate about the level of societal distrust in this type of initiative, there is recognition that with an understanding of the tragic consequences of nuclear weapons, there likely was an elevated level of concern related to any new nuclear technology. It would have been expected for society to seek assurance that these new technologies were safe.

Figure 7.2 shows a key table from the WASH-1400 study, aimed at evaluating risk for the peaceful use of nuclear research. The table suggests that the likelihood of early fatality caused by the failure of nuclear power stations was very low. The probability was then directly compared to other risks, with noticeably higher probabilities. The surprisingly low failure probabilities were the most noteworthy about WASH-1400 study results. The table only looks at individual risk but not risk related to large extreme events. This low probability supported the larger goals of the Atomic Energy Commission to expand the use of nuclear power. The report showed little transparency on how those probabilities were determined and examination of uncertainties.

In 1978, H.W. Lewis and others coauthored the Risk Assessment Review Group to the U.S. Nuclear Regulatory Commission, known as the Lewis Report. The report commended the attempts of WASH-1400 to advance risk assessment for nuclear reactors, using the following language:

> We do find that the methodology, which was an important advance over earlier methodologies applied to reactor risks, is sound, and should be developed and used more widely under circumstances in which there is an adequate data base or sufficient technical expertise to insert credible subjective probabilities into the calculations. Even when only bounds for certain parameters can be obtained, the method is still useful if the results are properly stated. Proper application of the methodology can therefore provide a tool for the NRC to make the licensing and regulatory process more rational, in more properly matching its resources (research, quality assurance, inspection, licensing regulations) to the risks provided by the proper application of the methodology. NRC has moved somewhat in this direction, and we recommend a faster pace.

TABLE 6-3   INDIVIDUAL RISK OF EARLY FATALITY BY VARIOUS CAUSES

(U.S. Population Average 1969)

Accident Type	Total Number for 1969	Approximate Individual Risk Early Fatality Probability/yr [a]
Motor Vehicle	55,791	3×10^{-4}
Falls	17,827	9×10^{-5}
Fires and Hot Substance	7,451	4×10^{-5}
Drowning	6,181	3×10^{-5}
Poison	4,516	2×10^{-5}
Firearms	2,309	1×10^{-5}
Machinery (1968)	2,054	1×10^{-5}
Water Transport	1,743	9×10^{-6}
Air Travel	1,778	9×10^{-6}
Falling Objects	1,271	6×10^{-6}
Electrocution	1,148	6×10^{-6}
Railway	884	4×10^{-6}
Lightning	160	5×10^{-7}
Tornadoes	118 [b]	4×10^{-7}
Hurricanes	90 [c]	4×10^{-7}
All Others	8,695	4×10^{-5}
All Accidents (from Table 6-1)	115,000	6×10^{-4}
Nuclear Accidents (100 reactors)	—	2×10^{-10} [d]

(a) Based on total U.S. population, except as noted.
(b) (1953-1971 avg.)
(c) (1901-1972 avg.)
(d) Based on a population at risk of 15×10^6.

FIGURE 7.2   Key findings from the WASH-1400 nuclear reactor safety study. Nuclear Regulatory Commission (1975). Obtained through Wikimedia Commons.

However, the Lewis Report also criticized the lack of transparency in calculations. It also found the Executive Summary to contain misleading language:

> Among our other findings are the well-known one that WASH-1400 is inscrutable, and that it is very difficult to follow the detailed thread of any calculation through the report. This has made peer review very difficult, yet peer review is the best method of assuring the technical credibility of such a complex undertaking. In particular, we find that the Executive Summary is a poor description of the contents of the report, should not be portrayed as such, and has lent itself to misuse in the discussion of reactor risks.

NASA and the nuclear industry learned heavily from applying probabilistic methods to understand and manage risk for their new and rapidly developing technologies. They also learned devastating lessons from serious incidents. The nuclear industry faced an intense need for innovating risk processes after the 1979 Three Mile Island accident at a nuclear station in Pennsylvania, involving a reactor failure that led to radioactive release. The incident revived interest in risk assessment for nuclear power. PRA grew in the nuclear field and became widely used in the oil and gas industry. Similarly, NASA experienced a tragedy with the 1986 Challenger Disaster, in which a space shuttle exploded within seconds of launch, killing all seven crew members. The disaster further solidified NASA's use of probabilities for risk assessment, using PRA, FMEA, and other tools.

FUKUSHIMA DAIICHI CASE – EVALUATION OF PROBABILISTIC RISK ASSESSMENT AT NUCLEAR PLANTS

While this chapter discussed the early development of PRA, the Fukushima Daiichi case involves a much more recent risk event. This risk event occurred well after the broad adoption of PRA in the nuclear industry but also shows the deficiencies in the use of PRA. In other words, adopting a single risk assessment tool is not necessarily effective without comprehensive evaluation of uncertainties, knowledge, effective decision-making, and a willingness to promote an appropriate risk culture.

Over the past many decades, there have been several calls for nuclear sites to refine protocols for earthquake risk further. Most notably, preceding the Fukushima Daiichi nuclear disaster, the Standards Committee for the Atomic Energy Society of Japan further clarified requirements for risk assessment methods, calling for all nuclear sites in Japan to review risk related to earthquakes in relation to new guidelines. In the case of Fukushima Daiichi, the plant had not yet completed this these activities.

PRAs for nuclear sites are generally classified using three levels: Level 1 (plant model) involves assessing risk of core damage by studying core damage frequency. Level 2 (containment model) involves assessing the release of radioactive material by considering both timing and magnitude. Level 3 (site model) involves assessing the broader impact of a risk event by considering deaths, injuries, and economic impact. Levels 1 and 2 are most common for nuclear sites.

The PRAs performed at Fukushima Daiichi included only internal event scenarios, including station blackouts. In addition, the scenarios did not consider a blackout involving multiple units, as was the case in March 2011. In addition, guidelines for PRA related to tsunamis were not available prior to the risk event.

There are many lessons learned after the Fukushima Daiichi nuclear event, and the authors suggest interested readers browse the full report by the National Research Council (2014), which was used for the details discussed in this case. Criticism of the risk event generally lies in the inadequacy of PRA methods when used in cursory ways. In this case, PRA was used for internal event scenarios and lacked sufficient consideration of external scenarios, such as those involving earthquakes and tsunamis. PRA was also criticized for its treatment of uncertainties, as the probabilities used lacked appropriate basis in uncertainties and knowledge used to inform those probabilities. The report includes the following two disadvantages of PRA (among many other disadvantages and advantages):

> PRAs that have been performed generally do not adequately account for human error in design, construction, maintenance, and operation of nuclear plants [...] or for intentional sabotage.
>
> The results of PRAs are limited by experts' ability to recognize all relevant phenomena, including potentially important external hazards, and by uncertainties and incompleteness of estimates of accident probabilities and consequences.

The nuclear site has also been highly criticized for being largely unprepared for the risk event, despite the use of risk assessment techniques. Additional disruptions slowed response efforts, including complications in access due to lack of power, damage, and aftershocks. The risk event occurred during a widespread risk event (earthquake and tsunami) that impacted other operations, not only the incident at the nuclear site. Thus, the infrastructure and emergency response systems were also burdened with other response activities.

Despite issues with PRA, probabilities remain a commonly used tool for representing uncertainties. However, the PRA discussion highlighted the need to consider the supporting knowledge, particularly the knowledge strength when assigning those probabilities.

Suppose one seeks to assess the probability of a severe nuclear event within the next year. There are large uncertainties due to the many factors influencing this probability, including the many discussed in the earlier example (e.g., natural disasters, concurrence of other disruptive scenarios, and other potentially unknown events). However, it could be relatively simple to assign a single number for that probability and make that number appear credible. For example, one could say:

> The probability of a nuclear incident is 0.0001%. This value is sufficiently low, so the risk is sufficiently low and tolerable.

The statement has some rationale if this probability is founded on a strong knowledge, using relevant data, information, modeling, testing, analysis, and argumentation. There is also an assumption that this knowledge is based on established methods and considerations that are outlined in relevant regulations and standards related to the industry. However, if the supporting knowledge is weak, it would not be justified.

To incorporate judgment about knowledge supporting the probability, an alternative formulation could be used, including:

> The likelihood of a nuclear incident in the next year is judged to be low, between 0.00001% and 0.01% based on the following knowledge [insert rationale for the knowledge basis...]. The supporting knowledge is judged to be relatively weak.

When making those strength of knowledge judgments, a variety of factors can be considered, including the amount and relevancy of data, information, and expertise considered. There is also consideration of the degree to which the phenomena involved are understood, as some topic areas have strong histories and scientific understanding while others do not. Similarly, there may be situations where accurate models exist, such that the accuracy has been examined across relevant situations. Another factor to consider is the degree of agreement among experts, as some risk issues may spark controversial judgments, skepticism, and general disagreement over scientific evidence.

There is particular concern over assumptions that are used to inform the strength of knowledge in uncertainty characterization. Assumptions may be made about some scope of events and not others (e.g., external vs. internal events), economic conditions, political factors, or other considerations. Assumptions could be made about elements that are largely out of the organization's control or domain knowledge. Knowledge strength can be rather weak without consideration of the dynamics, variability, or potential changes in those elements.

In the case of nuclear power and space travel, those were relatively new advancements that had only recently been operationalized between the 1960s and 1980s, thereby justifying weak knowledge. However, that weakness in knowledge is a relative judgment, such that the knowledge may not have appeared to be weak at the time. Similarly, the strength of knowledge can be determined by considering the accuracy of modeling with the data and information used. In the case of nuclear power or space travel, computational methods, data collection, and computational power were in very early stages in the 1960s and 1980s but may have appeared relatively strong at the time because they represented the current state of the art in scientific research.

The discussion of this chapter is strongly related to concepts of knowledge. Knowledge will be further discussed and explored in the following chapter.

WORKS CITED

Eisenhower, D. (1953). Atoms for peace speech. https://www.iaea.org/about/history/atoms-for-peace-speech

Lewis, H. W., Budnitz, R. J., Kouts, H. J., Loewenstein, W. B., Rowe, W. D., von Hippel, F. and Zachariasen, F. (1978). Risk Assessment Review Group report to the US Nuclear Regulatory Commission. [PWR; BWR] (No. NUREG/CR-0400). Nuclear Regulatory Commission, Washington, DC (USA). Ad Hock Risk Assessment Review Group.

National Research Council report Lessons Learned from the Fukushima Nuclear Accident for Improving Safety of U.S. Nuclear Plants, authored by the Committee on Lessons Learned from the Fukushima Nuclear Accident for Improving Safety and Security of U.S. Nuclear Plants. 2014 https://nap.nationalacademies.org/catalog/18294/lessons-learned-from-the-fukushima-nuclear-accident-for-improving-safety-of-us-nuclear-plants

Nuclear Regulatory Commission (2024). Probabilistic Risk Assessment (PRA). https://www.nrc.gov/about-nrc/regulatory/risk-informed/pra.html

Nuclear Regulatory Commission (1975). Reactor Safety Study, WASH-1400 – Table 6-3: Individual risk of early fatality by various causes. Original source: Reactor Safety Study, WASH-1400; Sourced through Wikimedia Commons. https://commons.wikimedia.org/wiki/File:WASH-1400_-_Table_6-3.png

FURTHER READING

Aven, T. and Thekdi, S. (2022). Risk Science: An Introduction. New York: Routledge. Chapter 3.

Flage, R., Aven, T., Baraldi, P. and Zio, E. (2014). Concerns, challenges and directions of development for the issue of representing uncertainty in risk assessment. Risk Analysis, 34(7), 1196–1207.

Lindley, D. V. (2006). Understanding Uncertainty. Hoboken, NJ: Wiley.

U.S. N.R.C. (2016). WASH-1400 the reactor safety study the introduction of risk assessment to the regulation of nuclear reactors. https://www.nrc.gov/docs/ML1622/ML16225A002.pdf

8 Types of Knowledge That Inform Understanding of Risk

Decoding Black Swans and Other Historic Risk Events

In June 2002, shortly after the 9/11 terrorist attacks, Secretary of State Donald Rumsfeld introduced and popularized the concept of "unknown unknowns." He explained the topic as follows:

> Now what is the message there? The message is that there are no "knowns." There are thing we know that we know. There are known unknowns. That is to say there are things that we now know we don't know. But there are also unknown unknowns. There are things we don't know we don't know. So when we do the best we can and we pull all this information together, and we then say well that's basically what we see as the situation, that is really only the known knowns and the known unknowns. And each year, we discover a few more of those unknown unknowns.
>
> It sounds like a riddle. It isn't a riddle. It is a very serious, important matter.
>
> There's another way to phrase that and that is that the absence of evidence is not evidence of absence. It is basically saying the same thing in a different way. Simply because you do not have evidence that something exists does not mean that you have evidence that it doesn't exist. And yet almost always, when we make our threat assessments, when we look at the world, we end up basing it on the first two pieces of that puzzle, rather than all three.

While Donald Rumsfeld was not the first to classify knowledge in this way, popularizing the terms inspired discussions and advancements in risk science. As discussed in this book, understanding and classification of knowledge are vital for characterizing risk.

Knowledge can take on many forms. To generalize, knowledge consists of justified beliefs. Those beliefs can be informed by evidence, as suggested in the preceding quote. That evidence includes data, information, modeling, and testing, argumentation. Across the case studies studied in this book, that knowledge includes aspects like information from news reports, email exchanges, mathematical models, data mandatory reporting, system monitoring, and statistical sampling with analysis.

DOI: 10.1201/9781003437031-8

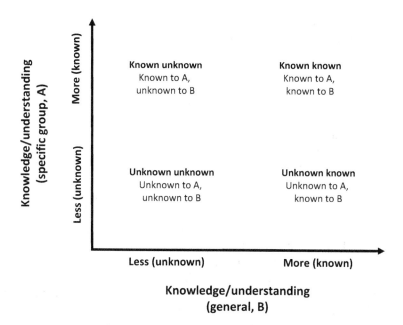

FIGURE 8.1 Classification of knowledge between some specific group A and general understanding B. (Based on Thekdi and Aven, 2023.)

That knowledge encompasses two distinct categories: General Knowledge (GK) and Specific Knowledge (SK). Suppose we classify knowledge related to the Flint water crisis case study. GK refers to generic knowledge, such as knowledge related to civil engineering (water treatment) principles, policies (e.g., Safe Drinking Water Act), and scientific knowledge about health consequences (e.g., lead contamination, bacteria). SK refers to knowledge about specific situations, such as the quality of the existing pipes, resident complaints, water quality test results, medical analyses, and the context of the discourse in email exchanges among officials.

Donald Rumsfeld's statements distinguish between general understanding/ knowledge of a specific group (A) versus general understanding/knowledge that is broadly known (group B), as shown in Figure 8.1.

Suppose we apply the framework of knowledge understanding between groups to the Flint water crisis case study. Specific group A represents water treatment staff, officials, and policymakers. General group B represents the general public, including residents, the media, and research groups. We can then classify the knowledge between groups into four distinct categories:

Known unknown (Known to A, unknown to B): Group A had detailed information about how the water was being treated. That knowledge included whether there was adequate corrosion control, their internal assumptions about pipe quality, and practices for water quality sampling (e.g., selection of sample data, whether to sample after pre-flushing the system). Group B had relatively less knowledge about these matters but had clues (e.g., taste, odor, and color).

Known known (Known to A, known to B): Both groups A and B had knowledge about water quality aesthetics. Records show that group B was vocal about these issues (e.g., taste, color, and odor). There were accusations that group A was largely dismissive about these complaints and assumed that these issues were specific to residential plumbing, not the water treatment system itself. There are also allegations that group B was given falsehoods, including being told that the witnessed water issues were "aesthetic" with no implications on health and safety.

Unknown unknown (Unknown to A, unknown to B): The long-term repercussions of water quality issues on behaviors were unknown to groups A and B. The Flint water crisis had significant media coverage, and it is that momentum to uncover the true state of water quality and the associated visibility of that effort that contributed to the risk issue being understood and addressed. In the years since the risk event, other cities have faced comparable risk events. However, it may not be apparent where hidden water quality issues remain. Similarly, water availability and quality remain a global problem, with billions of people lacking safe drinking water, with disastrous health repercussions. The long-term effects can also expand beyond the health aspects discussed here and include the environmental, socioeconomic, political, and justice/fairness impacts of the risk event.

Unknown known (Unknown to A, known to B): While the details of information flow are limited to email records that have been made public, it can be assumed that group A had limited knowledge about the state of water quality. Knowledge about this true state of water quality would need to be determined by using widespread independent sampling of households. However, external organizations (universities and healthcare industry) were able to perform comprehensive sampling by working with residents. This is an initiative that is often labeled *popular epidemiology*, because rather than the public sector, it was residents and those representing those residents who collected and analyzed the data. While group A has pre-determined sampling methods and assumptions, group B as a whole cohesive group has a wide array of data and information (e.g., water samples, taste, and color) that would not be available to group A (Figure 8.2).

It may also be alarming to see health and safety issues so low on the knowledge scale. As discussed in previous chapters, the precautionary principle would take precedence in these cases, calling for groups to take mitigative measures when faced with a potential serious threat when there are large scientific uncertainties related to knowledge and understanding of the risk.

The knowledge classification provides insights into priorities for improving knowledge within particular groups. Those in risk roles, presumably within group A (but not necessarily in all cases), may seek to have the most important knowledge issues move higher on the vertical axis. Those in the general public, presumably group B (but also not necessarily in all cases) may seek to have their most important risk issues move further right on the horizontal axis.

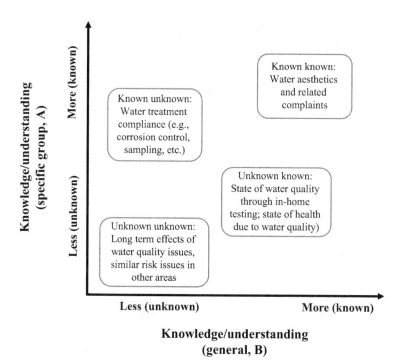

FIGURE 8.2 Classification of knowledge between water treatment staff, officials, and policymakers (group A) and the general public (group B). (Based on Thekdi and Aven, 2023.)

However, it is also not a simple task to determine what dimensions of knowledge matter. In the duration of a risk event, it might not be apparent what pieces of information are important (e.g., which water quality sample should be used for inference or which type of complaints are legitimate). That type of understanding might not become apparent until much later after the event has unfolded.

In addition, it might not be apparent which aspects of the knowledge are much more relevant or of high priority at any particular place and time. For example, was abiding by the interpretation of legal requirements (e.g., knowledge about the legal system, liabilities, and best practices) the highest priority during the Flint water crisis? Or were knowledge dimensions about long-term health, fairness, etc. of the highest priority? Those in turn, are value judgements that take place in the individual, institutional, and team setting. Given the many stakeholders involved, it is also likely that bureaucracy and disagreement on aspects of knowledge were contributors to the severity of the risk event.

We will further think about knowledge in the context of national security as it relates broadly to the 9/11 case as follows.

9/11 CASE – CLASSIFICATION OF KNOWLEDGE

Issues of national security strongly relate to the balance of knowledge between groups. While much of the discussion of this chapter simplifies knowledge among two distinct groups, areas of national security involve a much more complex web of knowledge transfer. This case study investigates the knowledge related to the precursors for 9/11 and the long-term impacts on knowledge gathering and use.

Prior to the 9/11 terrorist attacks, the United States had extensive knowns about related terrorist attempts directed toward United States landmarks. Less than 10 years prior to the 9/11 attacks, the February 1993 World Trade Center bombings caused six deaths and over 1,000 injuries. Within the parking garage rubble, officials uncovered evidence of a rented van used for the bombing. One of the co-conspirators was arrested upon his attempt to get his $400 deposit returned for the reportedly stolen van. In 1998, United States embassies in Tanzania and Kenya were bombed, resulting in over 200 deaths. Then, in October 2000, terrorists attacked the U.S. Navy ship USS Cole, killing 17 crew members. Intelligence analysis led to other arrests and the later uncovering of other ongoing plans for related terrorism campaigns.

The bombings began a new era of increased security. Also, there was increased attention toward emergency management and planning in the case of such attacks. However, the arrests made in response to the past attacks sent a signal that intelligence gathering and proactivity on behalf of government agencies were largely effective. There was also general disbelief that an attacker returned for a $400 truck deposit, suggesting that the attackers were not poised for large-scale complex attacks. Due to these factors, combined with bureaucracies, poor coordination and information sharing among agencies, and political interests, it can be argued that the United States mischaracterized the risk related to future attacks. This is likely a case of risk attenuation, which will be discussed in later chapters. Despite warnings from policymakers and FBI officials, intelligence agencies and law enforcement remain criticized for not doing enough to address risk related to terrorism threats.

In the months leading up to 9/11, there was indeed evidence of terrorist threats. This evidence was plentiful, as described in the 9/11 Commission Report (2004), which was used as a reference for this case. However, the intelligence signaling impending attacks was fragmented and sometimes classified. The President's Daily Brief (PDB) was used to inform the president about intelligence updates. It is estimated that between January 20 and September 10, 2001, over 40 articles in the PDB included topics related to the 9/11 attackers. One notable example of intelligence on the impending attacks was the Phoenix Memo of July 2001. In the memo, an FBI agent warned about potential terrorist efforts to attend aviation schools. The memo recommended intelligence gathering about persons attending those flight schools, including visa information. However, that memo was not widely read, distributed, or used to promote specific mitigation actions.

Donald Rumsfeld's famous quote was in the context of the aftermath of the 9/11 terrorist attacks. In the short term, there were many questions about the cause, magnitude, and source of the individual attacks. Some of the most critical questions included whether the attacks resulted from accidents versus orchestrated plans. Additionally, there were questions about whether future attacks were impending.

Much of the evidence that was used to answer those questions remains classified. Security and classification of available knowledge were paramount in longer-term efforts to address terrorist threats. However, action was swift. Beginning on the day of the 9/11 attacks, the Immigration and Naturalization Service and the FBI began the process of addressing immigration violations in response to leads about the attacks. The intent was "risk minimization" to acquire information about the past and potential future attacks.

Knowledge transfer to the general public was also of high priority. On the morning of 9/11, President Bush was scheduled to read to a class in Sarasota, Florida. During the event, the president was informed that a plane had crashed into the World Trade Center, potentially due to pilot error. After the second plane hit the World Trade Center, it was apparent that an attack was ongoing. By 9:55 a.m., the president had boarded Air Force One en route to find a safe location to deliver a public speech, eventually landing at the Barksdale Air Force Base in Louisiana. Figure 8.3 shows the original notes written by President Bush for his first speech to the press after the attacks on 9/11. The note suggests a careful selection of words to describe the attacks. The inclusion of the term *terrorist* signaled future stronger language about the impending war on terrorism.

As the terrorist attacks unfolded, phone communication among the White House staff, Defense Department, FAA, and other agencies was "of little value, and there were other important tasks," according to the 9/11 Commission Report. As a result, there was little coordination in response and decision-making activities. Phone calls continued throughout the day, but equipment problems and phone security hampered efforts. Most notably, there remained confusion over authority and the content of directives to shoot down potentially hijacked planes that did not respond to directions. There remained much confusion over the rules of engagement in the Washington, DC, airspace.

Days after 9/11, the USA PATRIOT Act was formed. The goal was to improve mechanisms for information sharing. In times of large intelligence gathering, it becomes difficult to separate and identify credible knowledge that can be effectively used for homeland security. That intelligence was also combined with large uncertainties (e.g., the magnitude of future attacks, the financial conditions of the terrorists, and who was harboring the terrorists).

The balance of knowledge also became more complicated in the years following the 9/11 terrorist attacks. Most notably, the use of the Internet and associated Internet communication tools like social media, 24-hour news feeds, and the propagation of bite-sized information have changed the way we communicate knowledge about risk issues to the general public. These new technologies have also further enabled malicious organizations to communicate, recruit, and propagandize.

The aftermath of 9/11 led to a "War on Terror." The campaign was widespread, with wars in Afghanistan and Iraq, and other global military operations. While discussions following the terrorist attacks had many elements, the most significant element was related to seeking evidence that other nations were providing weapons of mass destruction (WMD) to terrorist groups.

As part of the War on Terror, the United States and other nations drastically increased their intelligence-gathering operations. The United States National Security Agency was widely criticized for its data collection procedures, which involved sensitive data about nearly all residents. The agency was found to have kept data on

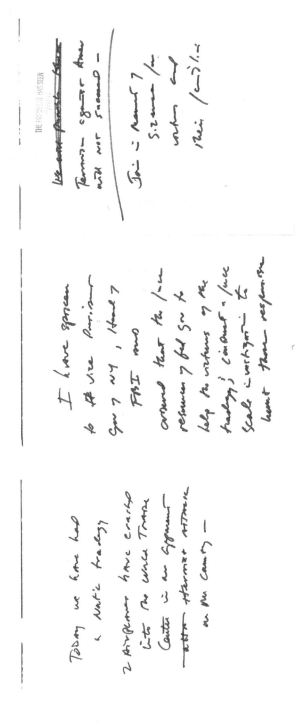

FIGURE 8.3   Notes written by President Bush on 9/11. National Archives (2001) photo no. 31490798.

phone call records, email exchanges, social media records, and Internet activity. The agency also had capabilities to collect more detailed data on specific people, including detailed email accounts and recorded phone calls, under the Foreign Intelligence Surveillance Act. There was a large degree of legal and civil controversy over these practices. Nonetheless, this provides an example of efforts to improve knowledge about factors that the United States *knew they didn't know.*

The increase in intelligence, combined with continued focus on eliminating major threats to the United States, led to knowledge of the whereabouts of the 9/11 attacks leader. In 2011, the United States Navy SEALS, on the orders of President Barack Obama, carried out a 40-minute raid of the Pakistan compound hiding the leader of the 9/11 attacks. Figure 8.4 is the famous Situation Room photo showing President Obama and other high-ranking national security team leaders receiving updates on the raid.

The use and spread of knowledge have created pathways for further study, monitoring, and analysis of security-related knowledge. For example, the United States Department of Homeland Security was created shortly after the 9/11 attacks. This agency was tasked with developing risk management frameworks and processes that promote coordination among government agencies, the private sector, and other parties. While the agency's work has been both lauded and criticized, there was a

FIGURE 8.4   President Barack Obama and other high-ranking national security team members receiving updates in the Situation Room of the White House on May 1, 2011, on the killing of the leader of the 9/11 attacks. Wikimedia Commons (2011). Photograph taken by Pete Souza. Original Source: White House Flickr Feed.

general step in the direction of knowledge-sharing across various stakeholders in attempts to improve awareness and systematization for the characterization and management of risk.

In more recent times, terrorism has become even more complex. There is also increased discussion about cyber-physical risk, relating to how cyberattacks can have physical consequences. Examples of cyberattacks with physical consequences could include attacks on healthcare facilities, autonomous vehicles, and the electric grid. With cyber-terrorism, the balance of knowledge related to threats has taken on technological dimensions.

Similar to the concept of popular epidemiology discussed in this chapter, the increased attention toward national security has further increased the role of citizens in aiding in risk management. There is increased attention on whistleblowers, or more simply citizens, reporting threats and other risk-related concerns. This reporting is a critical tool for knowledge transfer. The famous line "if you see something, say something" from the Department of Homeland Security adequately summarizes the importance of crowdsourcing key knowledge.

Relevant knowledge is obtained from several sources. Often, this comes from training, academic knowledge, rules and regulations, surveillance, monitoring, and other sources. Knowledge may also be sourced from those in risk communication roles, where there is a question about deciding on the detail, complexity, content, or tone of a risk message. As the 9/11 case study describes, communication can be of little value if it is not appropriately conducted. These issues will be further discussed in Chapter 10.

This chapter explored how knowledge is not just about informing the characterization of risk, but also that knowledge is in comparison to that of others. One does not necessarily need to seek all available knowledge across all aspects of the risk. Too much knowledge makes it difficult to identify what is meaningful and relevant versus what is not, creating confusion and indecision. Similarly, too much knowledge of varying credibility makes it difficult to identify what knowledge is credible and actionable within the perspective of also acknowledging associated uncertainties, as described in the 9/11 case study. Instead, a concerted effort to study knowledge is combined with judgment in identifying the highest priorities for knowledge collection. It also implies that one already has defined the risks' most important aspects (e.g., health, safety, adherence to policy, political image, and reputation).

That available knowledge, however, can have varying degrees of credibility. We will further discuss this in the following chapter.

WORKS CITED

9/11 Commission. (2004). The 9/11 Commission Report. https://9-11commission.gov/report/911Report.pdf

Rumsfeld, D. (2002). Press Conference by US Secretary of Defence, Donald Rumsfeld. https://www.nato.int/docu/speech/2002/s020606g.htm

National Archives (2001). Notes written by President George W. Bush for the initial statement to the press after the terrorist attacks on September 11, 2001. https://catalog.archives.gov/id/31490798

Wikimedia Commons (2011). The Situation Room. Photograph taken by Pete Souza. Original Source: White House Flickr Feed. https://commons.wikimedia.org/wiki/File:Obama_and_Biden_await_updates_on_bin_Laden.jpg

FURTHER READING

Aven, T. and Kristensen, V. (2019). How the distinction between general knowledge and specific knowledge can improve the foundation and practice of risk assessment and risk-informed decision-making. *Reliability Engineering and System Safety*, 191, 106553.

Thekdi, S. and Aven, T. (2023). *Think Risk: A Practical Guide to Actively Managing Risk*. London: Routledge.

9 Credibility of Knowledge

As discussed in the previous chapter, knowledge consists of justified beliefs. These beliefs can be informed by data, information, modeling, testing, and argumentation, which we will refer to as evidence for the purpose of this chapter.

There are three lenses to consider this evidence. First, a high-quality risk study is expected to leverage high-quality evidence, whether that evidence comes from internal or external sources. When referring to evidence from external sources, one would need to consider the trustworthiness of the evidence source and the methods that sources use to gather, analyze, and present the evidence. Internally, on behalf of the risk analyst, it's also imperative to aspire for high-quality evidence through credibility in data collection, analyses, and presentation of the results to others. While credible evidence is preferred, the reality is that the data source, analysis procedures, and the validity of arguments can all be questioned and have varying degrees of credibility.

The second lens is how individuals interpret the evidence. Individuals or groups can understand the evidence, misunderstand the evidence, or rely on others to generalize the findings. Two individuals can look at the same data and analysis and come up with different conclusions, potentially due to their own life experiences, values, contextual factors, biases, and perception-related factors. Sometimes, individuals want to hear what they want to hear (a form of confirmation bias), allowing them to only focus on evidence supporting their prior beliefs. Often, issues of skepticism and trust also impact that interpretation of evidence (recognizing that skepticism and trust can be justified). Additionally, any evidence is not necessarily rigid and fully factual. For example, consider statistical analysis in which there is some uncertainty to consider in the interpretation. All parties do not necessarily understand that uncertainty, and it's also not always clear how that uncertainty translates to understanding the risk issue being considered.

The third lens is how groups interpret the evidence. Interactions among group members, such as those related to power dynamics, group-think, and sociocultural factors, can influence how groups interpret information, make decisions, and act on information. In addition, due to the increased use of bite-sized pieces of information, there is recognition that not all group members will take the same amount of effort and interest in understanding and interpreting any presented evidence.

As we discuss the credibility of knowledge, we assume that individuals seek objective information about risk issues, leaving discussion of biases to later chapters. However, we also recognize that individuals may hesitate to change their minds or reverse stances on risk issues. That is problematic from a risk standpoint, but it's important for risk science research to recognize that habits and strongly held beliefs are not necessarily something individuals want to change, need to change, or should change, despite new or emerging evidence on various risk issues.

DOI: 10.1201/9781003437031-9

Let us first discuss what constitutes evidence. Evidence can come from data, such as surveys, photos, sensors, and transactions. While it's commonly believed that data is numeric or cleanly fits into spreadsheet cells, there are increasingly sophisticated tools to understand qualitative data (e.g., social media posts and book chapters). Any data should ideally come with "fine print" or metadata that includes elements like the author, time period studied, and assumptions made. Evidence can also come in the form of views and judgments, such as findings from scientific studies, expert opinions, and first-hand narratives of events. From an analytical perspective, evidence can emerge from the analysis of data, models, scientific testing, and hypotheses.

This evidence is used broadly across the understanding, management, and communication of risk. Using the work of Thekdi and Aven (2024), we consider the use of evidence in many aspects of a high-quality risk study:

Overall risk study: To conduct a high-quality risk study, the risk analyst would need to explore whether those conducting the study have a clear understanding of risk science fundamentals and have a clear risk question to be asked. For example, a risk study could be aimed at understanding the most critical vulnerabilities of an organization or could aim to characterize the risk related to some activity. Evidence could be used to monitor a risk study to ensure the study asks an appropriate risk-related question, whether the conducted study aligns with the study design, and determine whether individuals involved in the risk study are qualified.

Data and information: A high-quality risk study would require accurate information from reputable sources. The data and information would need to be applicable to the risk study and be representative of the main features of the risk study. For example, when studying risk related to working conditions (e.g., Dhaka garment factory fire), one would consider data and information about similar situations (e.g., similar laws, regulations, and conditions). Data and information are also processed in some ways. For example, it is common to clean the data, potentially removing outliers and handling missing values in the dataset. One would need to ensure that policies and assumptions for removing outliers and adding missing values are transparent, logical, and do not introduce systemic errors or biases in the risk study.

It can also be increasingly difficult to evaluate information to acknowledge the difference between fact and opinion, particularly when little basis, reasoning, or factual information is provided. Similarly, something prevalent in journalism and social media is the use of language that tends to minimize uncertainties, or, in other words, give the illusion of certainties when material uncertainties exist in the risk study. For example, consider an influencer on social media who offers weather predictions. That influencer can evoke certainties by guaranteeing a snowstorm, but from a scientific perspective, that guarantee is subject to uncertainties.

When data and information are collected, whether by the risk analyst or some external source, the quality of that data and information can be influenced by tangential factors. There may be industry standards for sampling data or deciding how to choose subjects for a study. The study itself can contain hidden biases. For example, consider a survey used to elicit information from stakeholders: the answers to the survey questions could be influenced by the phrasing, ordering of the questions, or scaling used for answers.

Finally, it's important to acknowledge and understand the implications of potential conflicts of interest. It should also be acknowledged whether the information has been discredited by third parties. Allegations intended to discredit do not necessarily invalidate the risk study in its entirety, but they need to be acknowledged and understood.

Analysis: There are many choices to be made when analyzing data, as illustrated by the following simple and well-known example: Consider an analyst seeking to find the average level of contamination for a public water source. The analyst could choose between computing the arithmetic average (adding up all measurements and dividing by the number of measurements taken) or using the median (the near-middle value in the sorted list of measurements). In the case of the arithmetic average, all measurement values are included in the computation, and the result could be strongly influenced by one or more very high/low values. When using a median, extreme high and low values are ignored.

There is also a question about the execution of an analysis. If a quantitative analysis procedure was used, it's important to ask whether it was used correctly and if the output of the analysis was correctly interpreted. It is not necessarily an easy task to ensure that all communicated outputs of an analysis are actually based on evidence from the study. To produce results, the analysts need to combine the evidence with their own judgments, when, for example, developing and using models. One should also be careful not to assume causation when relationships among factors are found within an analysis.

Managerial review and judgment, decisions, and communications: A risk study is never as simple as finding data, analyzing the data, and communicating the output. All models and analyses are simplifications of reality. That reality often involves many stakeholders with competing interests, a cultural environment, a social environment, economic conditions, political factors, power constructs, ethical concerns, stances on fairness, stances on equality/equity, and many other issues that are not simple. Thus, the outputs of a risk study are not simple and instead rely on considering many factors and weighing those factors with available evidence, values, and decision-making concerns. There may not always be a single right decision that meets all of the requirements and priorities for the risk study. Instead, a satisfactory decision will be made using the assumption of a high-quality risk study, sufficient evidence, transparency around risk study limitations, and a pre-determined decision-making process.

We will further explore the credibility of knowledge within the context of food additives.

FOOD ADDITIVES CASE – CREDIBILITY OF KNOWLEDGE

You may look around your office, kitchen, or grocery store and see a wide variety of colorful foods and personal care items. Some of those foods or care items may be colored white. For example, look closely at the packaging for coffee creamers, cake decorations, other candies, lipsticks, sunscreen, and toothpaste. Depending on where you live, you may find some of those products contain titanium dioxide, a food additive that allows products to appear whiter and which is used to extend the shelf-life of those products.

These types of food additives have a general purpose. They can be used to make foods appear more appetizing or fresher, make products look more interesting, and help patients distinguish between medications and other purposes. Food additives can also act as preservatives that decrease perishability, which can be particularly helpful in minimizing waste across the global food supply chain. However, these additives can also have severe health impacts, particularly as exposure to these additives can build up over time in a person's body.

While titanium dioxide is prevalent and highly studied, beliefs about its safety vary widely. Regulations over the use of titanium dioxide appear to be constantly changing. For example, one example of a regulation that was current at the writing of this book approved the use of titanium dioxide for the coloring of foods while subject to restrictions related to the quantity of titanium dioxide in reference to the percent by weight of the food. Another example of regulation banned titanium dioxide as a food additive. Similarly, other examples of regulations continue to allow titanium dioxide as a food additive.

Consider those regulations that ban titanium dioxide as a food additive. These types of bans are largely instituted due to the combination of policymaker values, scientific findings, uncertainties, and gaps in knowledge related to the additive's health impacts, or genotoxicity. The concerns over genotoxicity mean that there is potential for the food additive to damage DNA, potentially leading to cancers. Because studies could not conclude that the additive was safe, a precautionary approach could then be used to minimize exposure to the additive.

Broad evidence about the food additive may not be nuanced enough to clearly suggest a health risk for individuals. For example, reviews sometimes find different implications and evidence depending on whether food-grade titanium dioxide is studied. There may also be nuances related to the effects of the additive when combined with other elements of a person's diet, such as associated with the titanium dioxide particles binding to proteins.

One can generally ask: Why would there be so much disagreement about the safety of a product that scientists and food safety agencies have so highly studied? We explore this question by considering factors related to the credibility of knowledge.

While answering this question, the first thing to consider is whether the risk studies resulting in the policy decisions were of high quality. The first place to look is at the scientific research that informed those studies. While the authors of this book are not in a position to question the research methods and data considered by national panels of experts, the authors assume these panels consider an extensive review of literature and evidence. Thus, an external observer with risk training may reasonably conclude that the overall data, information, and analysis are of high quality. However, some of the studies may be more relevant than others. Thus, there may be some disagreement over the quality and relevance of data and information used to inform their respective risk-based decisions.

There is also a question of bureaucracy, decision-making processes, and related political factors. Given the governmental structures, decision-making processes are widely different across the world. There are widely varying time frames and processes for the review of food additives. One also cannot isolate policy-oriented decisions from the interests of lobbyists and other political interests with varying degrees of influence. Thus, one can conclude that other factors differentiating regulations are the management review and judgment, decisions, and communications.

At a more general level, one can ask how lawmakers consider uncertainties emerging from the conflicting evidence about titanium dioxide. One nation may choose to take a more precautionary approach when faced with uncertainties. Others may give different weight to those uncertainties. Thus, the judgments and decisions are to a large extent related to values.

There also appear to be differences among nations when making assumptions within their risk studies. For example, the literature suggests different assumptions when referencing the literature about food grade versus non-food grade titanium dioxide. There also may be differences in assumptions related to interactions with proteins in a person's diet. Similarly, there may be differing assumptions about dosage, as some regulation explicitly addresses percent by weight of food.

It should also be noted that time is an important element to consider in this case study. This case study was based on current regulations and knowledge available at the writing of this book. Regulation and general understanding of these issues evolve over time.

As discussed earlier in this chapter, there are three lenses to consider when evaluating knowledge. While this chapter has discussed the credibility of the evidence supporting the knowledge, the next chapter will discuss the other two lenses: How individuals interpret the evidence and how groups interpret the evidence.

FURTHER READING

Thekdi, S. and Aven, T. (2024). A classification system for characterizing the integrity and quality of evidence in risk studies. *Risk Analysis*, 44(1), 264–280.

10 Factors Influencing Understanding and Communication of Risk, Uncertainties, and Probabilities

Prior to the 1950s, the Weather Bureau in the United States banned the use of the word "tornado" in weather forecasts. There is speculation that this ban resulted from concern over emotional reactions to the word, such as fear, panic, and irrational behaviors when such events were predicted. This was also a time during which the field of meteorology was not equipped with modern data and analysis tools, so scientific understanding of tornadoes was in a very early stage. The poor abilities to forecast such storms may have contributed to unease with publicizing predictions. However, the dilemma at hand was how to handle the taboo of discussing these storms when the storms were responsible for tragic and potentially avoidable consequences across the country.

Figure 10.1 shows 1955 radar imagery of a tornado. While maps and tornado imagery in more modern times are much more graphic, it is understandable that these types of images evoked emotional responses. However, censoring words and the topic of tornadoes was also harmful from a risk perspective. As discussed throughout this book, clear communication and information are vital for appropriately managing risk. In this case, clear communication about the potential for severe weather combined with appropriate warnings that encourage residents to take protective action is essential for societal safety.

While the ban on the word tornado had understandable intentions, society has moved in the direction of being more upfront and communicative about risk. Communicating with appropriate words, imagery, and warnings allows those impacted by the risk to make informed decisions about understanding and responding to the risk issue.

In 1988, the United States Environmental Protection Agency published the Seven Cardinal Rules of Risk Communication (EPA 1988). While the methods of communication have changed drastically since the 1980s, the concepts behind those cardinal rules have longevity. While these rules may not be generally or consistently followed, the spirit of these rules provides meaningful insight into the communication of risk.

While these cardinal rules are phrased from the perspective of a governmental organization, the general concepts covered in these rules can broadly apply to any

DOI: 10.1201/9781003437031-10

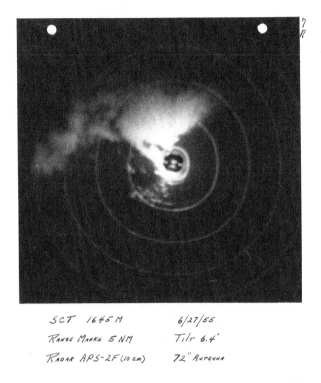

FIGURE 10.1   Radar imagery of tornado on June 27, 1955. National Archives (1955) photo no. 40570964.

organization with some level of power or influence over others. In more modern times, these concepts can apply to the media, influencers, corporations, and more.

We will discuss the basic concepts of these rules and discuss how those concepts relate to the historical cases discussed in this book:

EPA Rule 1: Accept and involve the public as a legitimate partner. This rule challenges a precedent of government officials and those in power attempting to prescribe the risk-based decisions of the public. This rule asserts the importance of having an informed public that should be trusted to participate in risk-based decisions, particularly when those decisions involve things they value (e.g., lives, health, property, and environment).

One complicating factor in promoting an informed public is the issue of uncertainties and knowledge. The reality is that many risk topics (e.g., climate change and new viruses) are not well understood. Thus, even the experts studying the risk topic may be building knowledge and informing themselves. As a result, many deficiencies in effective risk communication, including issues with misinformation (to be discussed in later chapters), emerge from communicating uncertainties and available knowledge, which are critical elements of risk characterization and understanding.

Even when uncertainties are understood and communicated, such as in the titanium dioxide example in Chapter 9, it is weighing of uncertainties, combined with values, that informs decision-making outcomes. In the case of broad community-level decisions, what is important is that the public is informed to the level of detail of understanding uncertainties and participates in the broad decision-making process. For example, consider decision-making related to removing lead cables in a community. Telecommunications companies historically used these cables, which pose serious safety risks to those in nearby areas. Ideally, the public should have a voice in deliberations about removing the cables and associated contaminants. In the case of individual decisions, suppose the public is informed about the existence of those contaminants in the soil and water. The public would need to understand the potential impacts of the contaminants on their health and the environment. They should also be able to use the available information, including information about uncertainties about location and levels of contamination, when deciding whether to expose themselves to the contaminated areas.

EPA Rule 2: Plan carefully and evaluate your efforts. This rule discusses the importance of taking risk communication seriously. Having a clear objective for the intent of risk communication can help guide efforts to know when to share a message and how to incorporate feedback into the message and promote dialogue.

This rule also highlights the importance of treating the public as a group of stakeholders that are not uniform in their concerns, priorities, values, and needs. For example, ideal methods of communication for various stakeholders may be different. Stakeholders receive communication in various ways, including television, email, text messages, landlines, and mail. The most effective communication method may vary significantly based on individual preferences and lifestyles.

An example of poor planning for risk communication involves communication of severe weather. In some areas, the local news television station will cut regular television programming to provide updates and warnings about severe weather situations, such as tornadoes. Concurrently, cable providers can cut into television programming to provide emergency warnings. Thus, the cable provider, who provides simplified warning messages, can stop viewers from accessing the more detailed messages. While both sources of risk-related communications intend to inform the public about the risk, the planning of such messages does not consider the possibility of competing messages that create confusion and block information in emergencies.

EPA Rule 3: Listen to the public's specific concerns. Interestingly, this rule states, "If you do not listen to people, you cannot expect them to listen to you." This statement appropriately highlights that risk communication is not just about saying but also about listening.

Gaining input from the public can be challenging. The EPA encourages the use of interviews, focus groups, and other types of listening exercises to gain public input. In more recent times, there is also an opportunity to gain input over social media, though this information may be of poor quality or be manipulated in some way.

This rule also makes a very important point about the needs and interests of the public versus those of the risk analyst. Those in risk analysis positions are often highly focused on data, information, statistics, and population-level factors. However, the interests of the public might not be so broad. The public may care more about the trustworthiness of the information source, fairness, values, and particular facets of the risk issue being discussed.

An example of the need to consider the public's specific concerns involves the COVID-19 pandemic. Throughout the pandemic, various authoritative figures, policymakers, and others promoted risk reduction measures intending to curb the spread of the coronavirus. Some of those measures were supported by available scientific knowledge, while there was also a policy component to the public health communication and decisions. However, individuals and groups also viewed those measures in coordination with other concerns, including aspects of freedoms, trust in science, interpretation of available scientific evidence, recognition of low knowledge of a newly discovered virus, distrust of government, conspiracy theories, and political narratives.

EPA Rule 4: Be honest, frank, and open. This rule emphasizes the importance of integrity when communicating risk issues. When expertise and influence exist for one party (e.g., the risk analysts or policymakers) and distrust or lack of information exists for the other party (e.g., the public or various subgroups), any risk communication attempts will be ineffective.

The rule also states the importance of honesty in communicating uncertainties and in acknowledging mistakes. The rule clearly and appropriately states that "trust and credibility are difficult to obtain. Once lost, they are almost impossible to regain completely."

As demonstrated in the tornado example, one major issue is censorship, which is a roadblock to honest, frank communication. When particular words or discussions about general risk topics are censored, society is less equipped to understand and manage risk. For example, consider an educational system that bans or discourages discussion about the severe risk events described in the case studies of this book. While the discussions of these risk events can be uncomfortable for teachers and students, these discussions further provide the tools for the next generation to think carefully about the risk in general and seek training for understanding and managing risk. Those individuals will be tasked with addressing generations of surprises (such as black swans) and risks emerging from constantly changing conditions. It is also those individuals who will improve upon the gaps in risk science that exist today.

Censorship can also exist in other harmful forms. For example, consider the practice of enterprise risk management, in which firms study potential risk events and further identify how to address the emerging high-priority risk issues. Teams can be reluctant to study the potential for a wide variety of risk events due to several factors. These factors include societal norms (e.g., hesitation to bring up uncomfortable topics), a culture of blame (e.g., blaming or inciting discord over past risk events), and hesitance to address risk issues that emerge, among many others. Censorship can also exist through content policies within social media, but that remains a complicated issue that is out of the scope of this book.

EPA Rule 5: Coordinate and collaborate with other credible sources. This rule states that, ideally, risk-related communications should be made in collaboration with other sources. There can be more than one trusted source, but agreement among those trusted sources can help reinforce messaging. The EPA's example calls for joint communications incorporating buy-in from credible scientists, physicians, and various trusted officials.

In reality, it can be challenging to manufacture collaboration among a wide range of sources. In more modern times, risk-related topics are highly politicized, making it difficult to identify whether agreements in messaging are due to political or scientific factors. Thus, any coordination and collaboration of risk-related messages must also have transparency, credibility, and accountability, which can be difficult to achieve.

EPA Rule 6: Meet the needs of the media. The EPA rule reminds us that the media is highly influential. The media's presentation of risk-related topics can be a key factor in the effectiveness of risk communication efforts. The rule calls for those in risk-related positions to be respectful and to maintain a good long-term relationship with the media.

The rule states, "The media are frequently more interested in politics than in risk; more interested in simplicity than in complexity; more interested in danger than in safety." While this statement was written in the 1980s, the media as it is defined today (e.g., social media and 24-hour news cycle journalism) appears to operate in similar ways. The politicization of risk-related topics and the financial incentives to create popular news stories incentivize entities to promote further political divisions on risk issues. There are also incentives to discuss danger versus safety, as themes of danger evoke emotional responses and panic, which can increase readership and influence. There is also an ideological component, as media sources can seek to be activists for an official view, promoting some dominant narrative on a risk topic area.

EPA Rule 7: Speak clearly and with compassion. This rule encourages the communicator to avoid "technical language and jargon" and generally be sensitive to the norms of the audience. Most importantly, the rule reminds us that risk topics can be highly sensitive, involving aspects of health and safety. Audiences may react to risk-related messages with feelings of anxiety, fear, anger, etc. The rule also reminds us audiences may react with disagreement or dissatisfaction, which is a normal component of risk communication activities.

The following example demonstrates the importance of risk communication as it relates to imagery.

CUYAHOGA RIVER FIRE – COMMUNICATION OF RISK-RELATED INFORMATION

While typical risk communication practices involve written and spoken words, there is recognition that communication of risk-related topics can be much more nuanced. Consider the imagery from one of the many Cuyahoga River fires in Cleveland, Ohio, USA, as shown in Figure 10.2.

Throughout the late 1800s and the early to mid-1900s industrialization, the Cuyahoga River became heavily contaminated with industrial waste, including debris, oil, and hazardous chemicals. While the condition was loathsome by today's regulatory norms, the pollution was also viewed as a sign of economic activity that promoted jobs and productivity.

The river experienced many fires over this time. One of the most publicly discussed fires occurred on June 22, 1969, when the industrial pollution ignited on the river's

FIGURE 10.2 1952 Cuyahoga River Fire. One of many fires on the Cuyahoga River during the time period. Photo by James Thomas. Cleveland Press Collection, Michael Schwartz Library. Cleveland State University. Digital archive no. 416. Cleveland Memory (1952).

surface. The highly publicized photo, similar in nature to the fire shown in the figure, was misattributed, as no records of the actual fire appear to exist. Instead, the photo was from an earlier fire. Nonetheless, the photo introduced citizens across the country to the impacts of environmental degradation due to industrialization, recognizing industrial pollution as a high-consequence and urgent risk issue.

The mayor of Cleveland, Carl Stokes, along with his brother, Congressman Louis Stokes, became known advocates for addressing risk issues due to pollution. The work was also motivated by the more broader issues of environmental justice encompassing how that pollution impacts aspects of equity, wellness, and citizen rights. They leveraged the publicity surrounding the fire and associated imagery to advocate for legislation and practices to address pollution and improve water quality. Figure 10.3 shows the mayor holding a press conference on the railroad bridge, presumably inviting the press to experience the sights, smells, sounds, and general perceptions of industrial pollution in the area.

The publicity from the Stokes brothers and many others encouraged policymakers to adopt stricter environmental regulations, leading to the passage of the Clean Water Act in 1972. This legislation includes requirements for industries to obtain permits for releasing pollutants into water bodies. The general momentum toward addressing environmental risk issues also led to stricter legislation addressing air and soil pollution. For example, the Resource Conservation and Recovery Act of 1976 improved hazardous waste treatment, storage, and disposal. Similarly, the Clean Air Act continued to expand over decades, signifying a growing awareness of risk related to air pollution, such as related to vehicles, smog, acid rain, and emissions from industrial sources.

Figure 10.4 shows a 1971 photo of a sign near a beach near Cleveland. The sign reads, "In the Spirit of… Cleveland Now Edgewater Beach Safe Swimming … Carl B. Stokes, Mayor." The sign emphasizes the importance of clean water not only for safety but also for recreation, community, and aesthetic aspects of the region.

FIGURE 10.3 Cleveland Mayor Carl Stokes speaking with reporters on a railroad bridge in 1969. Photo by Herman Seid. Cleveland Press Collection, Michael Schwartz Library. Cleveland State University. Digital archive no. 1337. Cleveland Memory (1969).

Indirectly, this sign publicly demonstrates the mayor's environmental justice values and provides rationale (e.g., recreation) for the community to share in the values of their elected leader.

As we consider the impact of the river fire imagery, we acknowledge the ability of a photo to share complex ideas, emotions, and thoughts. The photo sent different messages to different individuals, which were personal, political, and human. These messages were a form of risk communication as they illustrated the impact of a risk issue. Words alone could not have effectively expressed the message to policymakers, industry, and communities who rely on the industry, jobs, and the river. Thus, the Cuyahoga River Fire photo serves as a reminder for all seeking to communicate about risk-related issues:

First, risk communication is not just about data and information, and not simply about putting ideas out for the public to hear. Risk communication is also about people and things people value. Those things people value are personal, often relating to aspects of health, safety, and well-being. Each individual has history and experiences that guide their stances, understanding, and feelings about the risk topic.

FIGURE 10.4   Signage on beach in Cleveland: "In the Spirit of… Cleveland Now Edgewater Beach Safe Swimming … Carl B. Stokes, Mayor." Photo by Bill Nehez. Cleveland Press Collection, Michael Schwartz Library. Cleveland State University. Digital archive no. 1774. Cleveland Memory (1971).

Words, data, and statistics cannot always speak to these issues. Risk communication attempts must also be sensitive to these matters.

Second, the photo reminds us that communication is not only about words. Nonverbal cues, including imagery, sounds, formatting, and other sensory elements, are also vital in messaging. As Mayor Carl Stokes invited reporters to experience the risk issue firsthand on a railroad bridge, those reports and viewers gained a much more values-based understand of the risk issue.

Third, because risk is personal, risk communication is personal. When the risk message comes from a person, the message is perceived as coming from an individual, from lenses of past experiences, trustworthiness, cognitive factors, etc. In this case, the communication of risk was personal in that it was leveraged by elected leaders to address risk issues that affected many individuals. Perception of these messages must be considered when making a risk communication plan.

While the EPA's seven cardinal rules are extensively implemented and discussed, many more recent advancements in risk communication exist. One particular enhancement exists in the realm of emotional intelligence.

Research on emotional intelligence supports adopting practices to promote clear communication and trust, thereby enabling decision-makers to make rational choices that are not under duress. One frequently used communication practice includes deliberately using calmness to avoid triggering panic. In addition, promoting mutual respect among stakeholders is essential, especially in critical situations such as emergencies. These practices support the growing recognition that decision-makers have autonomy. Those decision-makers should be trusted to use their values and judgments without undue influence.

Following the topic of effectiveness in risk communication, there is also well-established research on trust and credibility from a general risk perspective. Ortwin Renn and colleagues provide a perspective on more granular factors contributing to credibility. The various dimensions of credibility not only exist within the message but also expand to the person who is communicating the message. It is also important to consider the institution, as that institution (e.g., a public agency, corporation, or non-profit) also carries a history and level of perceived trust. The communication is also to be seen in reference to the social climate, which is largely out of the control of a particular communicator or risk communication message.

Trust in leadership and in the communicator of risk messages can be helpful, particularly when that communication is based on high-quality risk analysis combined with honest communication. It is equally important to have a healthy skepticism that motivates the individual to seek additional information, compare risk messages across communication channels, and make risk-based decisions using available knowledge and individual values.

Much of the work of this chapter has focused on the perspective of the risk analyst communicating with the general public. However, another angle to consider is how the public perceives risk.

The basis behind the controversy over the ban of the word *tornado* is related to risk attenuation and risk amplification. In the late 1980s and early 1990s, Roger Kasperson and colleagues developed the Social Amplification of Risk Framework (SARF). This framework studies public reactions to risk events, considering many factors related to perceived risk. For example, filtering can influence perception, as humans only process some proportion of encountered information. Individuals also process information differently and apply their own social values to that information. Society, concurrence of other events, and other factors can also influence the perception of risk for individuals.

The SARF treats risk communication as signals that are both amplified and attenuated. Consider risk events that are judged to be relatively low based on available knowledge and information. However, suppose that there is a high degree of media coverage or open discussion about the risk issue, thereby causing the risk issue to be perceived as being relatively high. In this case, the risk is amplified. Conversely, consider a case in which the risk is judged to be relatively high based on available knowledge and information. Suppose the risk gains little media coverage or general interest from the population, thereby the risk is perceived to be relatively low. In this case, the risk is attenuated.

The SARF suggests that risk perception is influenced by both amplification and attenuation. Parties with influence over amplification include journalists, scientists, news and social media algorithms, politicians, celebrities, corporations, and others in positions of visibility and/or influence.

That perception of risk can also vary by the characteristics of the risk topic itself. We will discuss a few factors that can contribute to risk perception, recognizing that the response can vary considerably among individuals. These aspects have been well-studied by many risk researchers; the main conclusions will be summarized here. See the list of recommended resources at the end of this book for more information.

Dread: There tends to be a higher perceived risk when the consequences of the risk are dreaded or feared. Risk topics that involve dread include public health emergencies and pandemics such as the COVID-19 pandemic case discussed in this book, mainly due to their high mortality rates, potential for severe consequences, and other tangential consequences (i.e., social, economic, and political impacts). An example of a risk that may involve a lower level of dread, but is also of high concern, is the possibility of a major power outage. While a power outage can have disastrous consequences, that type of consequence is not typical when utility companies and backup systems effectively manage outages.

Trust: There tends to be higher perceived risk when there is limited trust in those institutions with control over risk exposure or management. These institutions could include government agencies, the media, financial institutions, or political parties. There may be valid reasons for that limited trust.

Voluntariness: There tends to be a higher perceived risk when risk exposure and consequences are involuntary. There are many examples of risk issues that are largely involuntary. For example, consider the case of forever chemicals in drinking water and other manufactured products. Facing societal concerns about chemicals, political bodies are making efforts to ban the chemicals. Several nations have developed ambitious goals including the language related to the phasing out of PFAS; however, the phasing out is balanced alongside the question of whether those chemicals are essential for a functioning society. Thus, the consideration of potential bans when considering scientific evidence also include other values-based criteria related to economies and meeting the needs of society. As a result, risk perception related to safe drinking water largely remains impacted by voluntariness issues.

Controllability: There tends to be a higher perceived risk when exposure and consequences are uncontrollable. For example, natural disasters, such as earthquakes, hurricanes, and tsunamis, are not controllable. The COVID-19 case study also demonstrated a lack of control over exposure to the virus in some settings, leading to masking and other interventions. Additionally, technological advancements, such as those related to artificial intelligence, may introduce unprecedented loss of control as tasks are delegated to the technologies.

Familiarity: There tends to be a higher perceived risk when those risk issues are unfamiliar. One example of unfamiliar risk issues is related to the impacts of climate change, as evidenced by the California climate crisis. There is also low familiarity of climate-related events' long-term and delayed impacts, such as implications for disease, agriculture, food insecurity, equity/equality, and biodiversity.

Media attention: Similar to the dynamics discussed in the SARF, risk issues that receive more media attention are more visible and are potentially more concerning to the public. The most visible risk-related discussions may not be fully objective or even true from a scientific and risk perspective. Nonetheless, it is the public interest in a risk story that can serve as income generation for many parties involved with sharing and managing content. In addition, significant risk issues may not receive a level of attention proportional to the risk priority. Examples include risks related to slow-onset environmental degradation, health conditions (e.g., heart disease or diabetes), and food and water insecurity. There is sometimes also a tendency to share a single viewpoint on these risk issues, when in reality, these risk issues can be very complex and involve many dimensions (as we discuss throughout this book).

In addition to risk communication and perception issues, there are more extensive levels of concern over general biases and distortions of risk-related information. The next chapter will discuss those topics.

WORKS CITED

Cleveland Memory (1952). Cuyahoga River fire, 1952 – Jefferson St. and W. 3rd. https://clevelandmemory.contentdm.oclc.org/digital/collection/press/id/605/rec/7

Cleveland Memory (1969). Mayor Carl B. Stokes holds a press conference on a railroad bridge. https://clevelandmemory.contentdm.oclc.org/digital/collection/press/id/8944/rec/142

Cleveland Memory (1971). Sign near Edgewater Park beach advertising safe swimming. https://clevelandmemory.contentdm.oclc.org/digital/collection/press/id/13239/rec/192

EPA. (1988). Seven cardinal rules of risk communication. https://archive.epa.gov/publicinvolvement/web/pdf/risk.pdf

National Archives (1955). 1645M Radar of Tornado at Scottsbluff, NE. https://catalog.archives.gov/id/40570964

FURTHER READING

Balog-Way, D., McComas, K. and Besley, J. (2020). The evolving field of risk communication. *Risk Analysis*, 40, 2240–2261.

Boissoneault, L. (2019). The Cuyahoga River caught fire at least a dozen times, but no one cared until 1969. *Smithsonian Magazine*, 19(06). https://www.smithsonianmag.com/history/cuyahoga-river-caught-fire-least-dozen-times-no-one-cared-until-1969-180972444/

Covello, V. T., Slovic, P. and Von Winterfeldt, D. (1986). *Risk Communication: A Review of the Literature*. https://www.researchgate.net/profile/Paul-Slovic/publication/285817518_Risk_communication_A_review_of_the_literature/links/5d3cd952a6fdcc370a6609e3/Risk-communication-A-review-of-the-literature.pdf

Fischhoff, B., Bostrom, A. and Quadrel, M. J. (1993). Risk perception and communication. *Annual Review of Public Health*, 14(1), 183–203.

Kasperson, R. E., Renn, O., Slovic, P., Brown, H. S., Ernel, J., Goble, R., Kasperson, J. S. and Ratick, S. (1988). The social amplification of risk: A conceptual framework. *Risk Analysis*, 8, 177–187.

Kasperson, R. E., Webler, T., Ram, B. and Sutton, J. (2022). Editorial. *Risk Analysis*, 42, 1367–1380.

Renn, O. (2008). *Risk Governance*. London: Earthscan.

Siegrist, M. (2021). Trust and risk perception: A critical review of the literature. *Risk Analysis*, 41(3), 480–490.

Siegrist, M. and Árvai, J. (2020). Risk perception: Reflections on 40 years of research. *Risk Analysis*, 40, 2191–2206.

Slovic, P. (2010). *The Feeling of Risk. New Perspectives on Risk Perception*. London and New York: Routledge.

11 | Biases, Misinformation, and Disinformation

In addition to the risk perception issues discussed in the previous chapter, larger issues exist in the individual understanding of risk. The first issue relates to biases that exist in how individuals gauge the severity of risk. The perceived severity of a risk can also be impacted by how information is presented, as the information can be of high quality or of poor quality leading to suspicions of falsehoods. We categorize those falsehoods as misinformation and disinformation.

There are several well-known and widely studied biases that are barriers to effective risk communication. These risk-related biases refer to cognitive tendencies or heuristics that impact how individuals use judgment to understand risk issues and subsequently make risk-related decisions. The implications for these biases expand across societal factors, technology, psychology, politics, and beyond. Some of the most highly studied biases in risk science literature include:

Availability bias: This bias occurs when an individual's probability assignment of a risk event is based on how easily they can recall other examples of conditions related to the risk topic area. When recent or highly impactful historical events are more easily remembered and recalled, individuals may perceive or judge the likelihood of a similar risk event higher than if such events are not easily remembered or recalled.

For example, consider the situation of industrial accidents such as the Seveso disaster or the BP oil spill. The wide-scale impact of these incidents can make them more memorable and easily accessible in people's minds. As another example, consider the case of the East Palestine train derailment. While statistics show that railway safety has improved drastically over the past several decades, the occurrence and memory of this specific risk event may lead many to assume that railroad safety has not been improving. As a result of availability biases, individuals may overestimate the frequency of these types of accidents due to the availability of highly publicized or widely reported incidents.

Representativeness bias: Similar to availability bias, representativeness bias occurs when an individual assesses the likelihood of a risk event by comparing the event to stereotypical numbers, such as based on previous experiences, information, and knowledge. The probability assignments based on those previous experiences, information, and knowledge may neglect to consider other aspects of credible knowledge.

For example, consider the risk related to PFAS. Representativeness bias may lead individuals to assess the likelihood of health repercussions based

DOI: 10.1201/9781003437031-11

on experiences with other types of unrelated contaminants. Those other types of contaminants may be better understood or differently regulated and therefore not reliably representative of PFAS risk level.

Anchoring bias: Anchoring bias involves a tendency to focus on initial information or reference points when making risk-related judgments. Individuals can anchor their risk judgments around some basic level that is informed by early opinions or past experiences with similar risk issues, and then adjust those judgments when new information emerges. For example, consider anchoring bias in industrial safety. If a company has a good safety record, with no major accidents for an extended period, individuals may anchor their perception of the company's safety performance to this high demonstrated safety record. Individuals may believe that the risk related to future incidents is low, despite insufficiencies in data, changes in operations, new threats, or evolving conditions.

Over time, anchoring bias can lead to complacency and insufficient critical study of risk related to various activities. The bias can lead to various parties, such as decision-makers and policymakers, to not practice due diligence or thoroughness in risk assessment, risk management, and emergency preparedness. For example, these parties may assume that lack of data, historically good performance, or some dominant narrative surrounding a risk issue provides a signal that the risk is appropriately understood and managed. This type of bias may have been a factor in the Fukushima Daiichi nuclear disaster case study: since past risk events of similar magnitude were not witnessed, preparation for large magnitude seismic risk events was not adequately considered in risk management initiatives.

Conversely, anchoring bias can lead to an opposite outcome. If a company has experienced a high-profile and high-consequence incident in the past, individuals may anchor their perception of the company's risk characterization to that past performance. This perception of a higher level of risk could persist even if substantial improvements and safety measures are implemented.

Confirmation bias: Confirmation bias refers to the tendency to seek, remember, and leverage risk-related information in a way that confirms preexisting beliefs. When characterizing risk or making risk-based decisions, individuals may focus on information that supports their existing or preferred views. In turn, individuals may not seek or leverage evidence that conflicts with those views. This bias relates heavily to the Titanic case, as many believed the ship was unsinkable, and designers therefore considered having a sufficient number of lifeboats to be unnecessary.

Confirmation bias is prevalent when applied to technology settings in which social media and search engines curate visible content in ways that cater to user interests and historical behaviors. Consider the example of confirmation bias as it relates to titanium dioxide used as a food additive, as discussed in the previous chapter. Individuals with preexisting beliefs about the safety of titanium dioxide may engage with research, media, or social groups that align with their preexisting beliefs. These individuals may also ignore or seek to discredit viewpoints that disagree with their preexisting beliefs. In addition, individuals may interpret inconclusive evidence or evidence demonstrating uncertainties, such as scientific studies, in a way that confirm their stances. Conversely, they may attribute information that conflicts with their stances to bias, flawed methodology, conspiracy theories, or other attempts to manipulate or misinform the public. Over time, confirmation bias combined with social settings in which individuals seek out others with similar beliefs can further validate and reinforce a particular belief.

Optimism bias: This bias refers to the tendency to assign low likelihoods to negative risk events when they are in reference to oneself and assign a relatively higher likelihood to negative risk events when they are in reference to others. Consider the optimism bias related to the Titanic disaster. The designers and others involved with the Titanic were overly optimistic about the sufficiency of the safety equipment (e.g., lifeboats and other design features). These individuals may have overestimated the ship's ability to withstand a collision with an iceberg and held sentiment like "the ship is its own lifeboat." In addition, the ship was traveling at a high speed despite being in an area known for icebergs. Optimism bias may have created an excessively low prediction of likelihood of hitting icebergs and a belief that the ship's advanced technology and precautions would prevent any serious accidents.

Publication bias: Publication bias refers to the tendency for particular qualities of risk-related research to dictate the likelihood of publication and dissemination of research findings. For example, academic journals may be more likely to publish academic papers when they have a particular theme that is within the scope of the journal, have findings that align with some dominant narrative or established belief, have findings that disprove some dominant narrative, and/or produce statistically significant results. Conversely, studies without those qualities can be less likely to be accepted for publication in a particular journal. As a result, publication bias can lead to overrepresentation of studies reporting some particular risk-related finding or narrative, potentially influencing the understanding of risk for a particular issue. As a result, as some ideas and findings are not published or made widely available to the public, the public can have an incomplete understanding related to the risk topic area. At the policy level, publication bias can result in risk-related decisions being based only on those findings that reach publication and are broadly influential after publication, which can potentially provide an incomplete perspective on the risk topic area.

Single action bias: Single action bias refers to the tendency to believe that a single action, initiative, or behavior, whether large or small, is sufficient to address a particular risk-related issue. Sometimes, that single action can be highly visible yet have a limited impact in relation to a complex and broad risk issue. This type of bias is prevalent across risk applications. The most common encounter of this bias is in the use of Enterprise Risk Management regimes in which organizations believe that box-checking and completing mandatory reporting are sufficient for addressing risk-related issues. Checking off boxes without actively understanding risk or creating a risk management culture is often to blame when large risk events happen in organizations with established risk programs. This single action bias can result in characterizing and managing risk in ways that implicitly ignore holistic aspects of the risk, such as basic knowledge, uncertainties, and other systemic issues related to the risk topic area. As a result, the bias leads to decision-making that may neglect appropriate mitigative actions. Essentially, this bias results in a false sense of accomplishment in understanding and managing risk.

In addition to the biases discussed in this chapter, moral hazards present other phenomena that influence or potentially mislead in the understanding and management of risk. A moral hazard is a situation in which the decisions and behaviors of an individual, organization, or other entity are influenced by the assumption that they will not bear the full consequences of those decisions and behaviors. A common example of a moral hazard is a football player who exhibits unsafe behaviors when wearing additional protective equipment, such as a helmet, thereby increasing the potential for serious injury.

On a larger scale, a moral hazard relates to an individual, firm, country, or other actor undertaking high-risk activities because they know some other entity will at least partially be responsible for the consequences of that risk event (if it were to happen). The moral hazard is incentivized when one is insulated from the potential adverse consequences of their decisions, leading them to act in ways they would otherwise avoid. Consider the insurance industry, in which the insured party is financially buffered against losses. The insured party may subsequently engage in particular behavior, recognizing that they are protected from the full financial consequences of that behavior. For example, if an individual has car insurance, they may engage in more dangerous driving behaviors, knowing that their insurance will cover the costs of potential accidents. As another example, consider using personal protective equipment (PPE) in an industrial setting. The use of PPE is critical and is highly regulated. However, workers may exhibit a moral hazard when they use PPE but fail to follow other safety protocols in the belief that wearing PPE is sufficient to cover all aspects of risk for the work activity.

The following case will describe biases and moral hazards related to the East Palestine train derailment and other similar freight accidents.

BIASES AND MORAL HAZARDS – EAST PALESTINE TRAIN DERAILMENT AND SIMILAR FREIGHT ACCIDENTS

The freight industry has relied on railway transportation since the mid-1800s. Rail transportation has undergone significant improvements over many decades. The industry has adapted to safety-oriented regulations and practices related to infrastructure inspection and the use of advanced technologies, including GPS, wireless communication, predictive maintenance, automated signaling systems, and many others.

The technology of concern related to the East Palestine train derailment was hot box detectors. These systems identify overheating bearings, known as "hot boxes," on locomotives. A hot box occurs when the bearing of a train wheel becomes excessively hot, potentially leading to serious failures. Hot box detectors are typically installed along rail lines at regular intervals, allowing operators to monitor the temperature of passing train wheels and bearings. When an abnormally high temperature is detected, the hot box detector should send an alert.

Detailed records of the testimony for the East Palestine train derailment, which were used as a reference for this case study (NTSB 2023), show that the hot box detector recorded increasing bearing temperatures. At the same time, there is conflicting information about alerts sent to the crew. Because the federal government did not regulate hot box detectors, there was no required spacing of these devices or required policies related to alerting crews and responding to alerts.

While this book does not make any claims about the legal, operational, or ethical implications of the policies surrounding the hot boxes in this particular case, the incident brings to light moral hazards that can emerge from overreliance on technology or, in some cases, insufficient human supervision for the use of safety-oriented technologies. There can be a tendency to leverage and trust those safety-related technologies without devoting sufficient attention to safety culture and precautionary approaches. As a result, organizations can overly rely on those technologies and be complacent in their risk regimes. Suppose operators and controllers become overly dependent on the technology and the associated assumptions used by the technology (e.g., spacing of hot box detectors, temperature thresholds for alerts, and communication methods for those alerts). In that case, there is reason for concern about reduced vigilance and a lack of active monitoring of critical factors such as track conditions, bearings, and other hazards along the route. In addition, overreliance on automated systems may result in operators receiving inadequate training on manual operation, emergency procedures, or critical decision-making, resulting in decreased levels of the ability, judgment, and intuition needed to respond to unforeseen circumstances or when technology fails.

In addition to perception-related issues, there are risk-related implications for falsehoods that exist within existing knowledge of communication. We will discuss those falsehoods with topics of misinformation and disinformation.

Within any scientific setting, there is often intent to be objective and stick to facts. Raw data is often interpreted to be value-free or fully objective. However, even raw data contains some assumptions, aspects open to interpretation, or inherent errors, particularly when considering how data is selected and used. Consider data that is as simple as a temperature reading for locomotive equipment. A scientific instrument takes that temperature reading with some intrinsic measurement error. The temperature reading can also be influenced by environmental factors, types of equipment being studied, the calibration process, and other factors that may not be apparent to system users. There are even more assumptions and nuances involved with other types of measurements. For example, surveys are particularly problematic as answers to questions can vary significantly depending on how questions are asked, the words used the order of questions, and many other subtle factors.

In recent years, there has been a lot of conversation about *misinformation, disinformation*, and *malinformation*. We categorize those terms as falsehoods that are maintained and shared with others. We classify misinformation as erroneous, inaccurate, or misleading information about a risk issue when the information sharer does not intend to deceive. An example of misinformation is spreading rumors that, unknowingly to the speaker, are false. We classify disinformation as falsehoods shared with the intent to deceive. Examples of disinformation include spreading rumors that are knowingly false to the speaker. There could be many motivations for disinformation, such as financial gain or political power. Disinformation can be prevalent, such as false claims within product advertising, spreading false information to encourage social unrest, or altering safety records to avoid legal repercussions. We classify malinformation as misinterpreted information, such as misinterpreting the results of a mathematical model used for risk assessment.

History has shown that risk analysis, risk management, and risk communication initiatives have been accused of being misinformation and disinformation. We compare misinformation and disinformation using several lenses, using the ideas described by Thekdi and Aven (2023).

First, consider the characteristics of information used for a risk study. A risk study can be viewed as misinformation if rules/standards are unclear, assumptions are contested, the risk approach is not effective to address the risk problem, or the computations have poor accuracy due to uncertainties. Disinformation can emerge if the risk study intentionally does not comply with established rules, the assumptions are not defendable or disclosed, there are conflicts of interest, irrelevant information is intentionally used as a basis for the risk study, or if computations are manipulated to support particular outcomes. It may be difficult to gauge whether a risk study presents high-quality information, misinformation, or disinformation in real risk applications. For example, a study could involve a risk issue in which there are conflicting rules, if judgments are not agreed upon by various stakeholders, if there are conflicts of interest related to funding of the risk study, or when there are high uncertainties and poor knowledge involved with the risk issue.

Second, consider the characteristics of evidence used for analysis. A risk study can be viewed as misinformation if there are gaps in the qualifications of experts, biases in expert elicitation, or gaps in data accuracy, applicability, or integrity. A risk study can be viewed as disinformation if experts have conflicts of interest or the risk study knowingly uses poor-quality evidence. In real risk applications, there may be accusations of misinformation and disinformation if there are undisclosed biases or if there is use of imperfect data containing measurement errors and other inaccuracies.

Third, consider the characteristics of any computational analysis. A risk study can be seen as misinformation if the selected analytical approaches are not based on risk science knowledge, the selected approaches are misleading or conflict with risk science principles, or if qualified professionals cannot verify the reading of the analysis. The risk study can be seen as disinformation if the analytical approaches are intentionally not appropriate or credible, if there are intentional errors in implementation, or if the analysis does not support the interpretation. In real risk applications, there may be accusations of misinformation and disinformation if there is no clear best approach/model, simplifications are made, or judgment is needed to interpret the analysis.

Fourth, consider the characteristics of judgment, decisions, and communications. A risk study can be viewed as misinformation if there is a lack of transparency in decision-making or if value-laden language is used by those communicating the results of a risk study. A risk study can be viewed as disinformation if a predefined decision-making process is not followed or if communication practices are intentionally deceiving. In real risk applications, there may be accusations of misinformation and disinformation if there are divergent stakeholder concerns, if decisions do not appear fair, or if there is a lack of transparency in communications.

It is important to remember that risk is about values. Not all individuals will share similar risk characterizations or value-based risk decisions. That lack of consensus does not necessarily suggest that a risk study is wrong or of poor quality. However, it is often typical to find that the scientific interpretations from qualified experts are similar.

The following COVID-19 case study will demonstrate how risk studies with the intent to be of high quality can be compromised in high uncertainty and low knowledge situations. The wide politicization of a risk topic area combined with misinformation and disinformation campaigns severely impacted abilities to coordinate effective mitigation decisions.

COVID-19 MISINFORMATION AND DISINFORMATION

Throughout the COVID-19 pandemic, there was intense discussion about misinformation and disinformation. The disinformation was widely studied as allegedly deceiving information spread on social media, various journalism websites, messaging platforms, and other types of social interactions. One example of disinformation involved dangerous and potentially deadly suggestions for cures and treatments for COVID-19, such as those involving ingestion of household chemicals. Other theories claimed 5G networks had a role in how the virus spread, and others claimed that COVID-19 vaccines included microchips intended to control individuals. Intense discussion remains about disinformation being sourced from authoritative figures.

Regardless of the source of the disinformation, there is recognition that it involves the intent to deceive or promote false information, which is inappropriate from a risk science perspective. At a larger scale, COVID-19 disinformation had the potential to create confusion about the severity of the pandemic or misrepresent scientific knowledge related to the risk. These issues can ultimately undermine public trust in risk studies, decision-making, and communication. Conversely, a level of skeptical trust in risk studies and communication can be vital for promoting active discourse around risk issues.

Various groups may have had inherent incentives to create and spread misinformation and disinformation. Individuals could share information through conversations, social media, and other communication mediums. Those individuals may themselves have encountered general falsehoods and repeated them to others and they may or may not have been aware of the inaccuracies involved. The falsehoods may have also aligned with individual personal agendas, confirmation bias, ideological beliefs, or financial incentives to promote certain narratives. Various actors may have intended to spread propaganda, shape narratives, and promote distrust or conspiracy theories. Political organizations or interest groups may have used misinformation to discredit opponents or manipulate public sentiment.

In addition, parties may have used bots or other technological tools to amplify false narratives or misleading information about COVID-19 by promoting, sharing, or commenting on social media posts. By boosting the engagement of specific narratives, those bots could influence their perceived importance or the public acceptance of the narrative, even if those narratives were based on what the scientific community deemed as falsehoods. The bots could also amplify conflicting viewpoints, contribute to online arguments, and create confusion and uncertainty. In general, widespread disinformation can further make it harder for users to identify manufactured content.

As a result of general distrust over information and policies, many risk studies and risk communication activities were labeled as misinformation and disinformation by various parties. The growing sentiment of distrust and polarization serves as a reminder that many narratives may exist surrounding a risk issue. Effective risk communication by qualified experts may do little to change the prevailing narratives. Competing narratives can be expected to exist in situations with high uncertainties and low knowledge. Those competing narratives can signal a healthy scientific process to gain further knowledge. However, competing narratives containing misinformation and disinformation can also severely compromise risk mitigation efforts.

There is no clear answer for how the risk science field can or should further address the issues of misinformation and disinformation. As this book discusses, risk issues are incredibly complex, with many dimensions and competing objectives. Choosing or promoting a single narrative can be problematic when competing narratives are silenced. The controversies over the competing narratives can be as harmful as the risk issue itself because of the potential to create divides among various groups, promote distrust, or otherwise promote dangerous behaviors.

One can ask if risk science can guide potential solutions to the issues of misinformation and disinformation. If risk science promotes an informed public that can use their values to make risk-based decisions, competing narratives surrounding risk issues can be a standard component of an informed society. However, there is vagueness in identifying misinformation and disinformation because nobody can say which narrative is more "truthful." However, we can look at other practices discussed over the past few chapters. There is a call for transparency and integrity in creating narratives and risk-informed conversations. The risk science profession serves as a leader in this domain because it promotes transparency, integrity, and ethics, which then become expected by decision-makers. As the risk science field grows, so does the public's expectation for ethical communication. These practices can help address future issues with misinformation and disinformation to an extent.

WORKS CITED

NTSB. (2023). Project summary: Rail public hearing – 279 docket items – DCA23HR001. https://data.ntsb.gov/Docket?ProjectID=106864

FURTHER READING

Aven, T. and Thekdi, S. (2022a). On how to characterize and confront misinformation in a risk context. *Journal of Risk Research*, 25, 11–12.

Aven, T. and Thekdi, S. (2022b). *Risk Science: An Introduction*. New York: Routledge. Chapters 8 and 9.

Montibeller, G. and Von Winterfeldt, D. (2015). Cognitive and motivational biases in decision and risk analysis. *Risk Analysis*, 35(7), 1230–1251.

Thekdi, S. and Aven, T. (2023). *Characterization of Biases and Their Impact on the Integrity of a Risk Study*. Safety Science. https://www.sciencedirect.com/science/article/pii/S0925753523003181

Tversky, A. and Kahneman, D. (1974). Judgment under uncertainty: Heuristics and biases. *Science*, 185, 1124–1131.

Slovic, P., Fischhoff, B. and Lichtenstein, S. (2016). Cognitive processes and societal risk taking. In *The Perception of Risk* (pp. 32–50). London: Routledge.

12 Balancing Various Dimensions of a Risk Application

As the cases of this book demonstrate, risk applications consider many different dimensions. The most commonly studied dimensions include health and safety, reputation, regulatory concerns, legal liabilities, environmental impact, and monetary profit and costs. However, as we discuss in this chapter, these studied metrics can be difficult to quantify. When they are quantified, they could contain many inherent assumptions. For example, assumptions could involve an engineering-based understanding of the system or the sufficiency of regulatory requirements in addressing the risk.

After the risk has been characterized, there is also a challenge in using the understanding of risk to guide decision-making. We remind ourselves of the options generally available to address a particular risk: accept the risk, transfer the risk, mitigate the risk, or avoid the risk.

While values guide decisions for addressing risk, the reality is that treating some risks can be a substantial investment of time, money, and resources. Due to these concerns, various practical tools have been developed in the finance, accounting, and engineering domains. Some popular mechanisms are cost–benefit analysis, used to compare monetized costs and benefits, and multi-attribute decision analysis, used to aid in decision-making when considering the risk issue across various dimensions. While specifics of those tools are out of the scope of this book, we can generally discuss the dimensions considered by those tools.

The value of a statistical life (VSL), also discussed in Chapter 6, often has a role in risk applications that aim to balance various concerns and to help inform the most appropriate risk-related decision for a given situation. VSL assigns an economic value to the reduction in mortality risk, representing the amount of money society is willing to pay to prevent a statistically expected loss of life. By attaching a monetary value to this reduction in mortality risk, VSL can aid in comparing the effectiveness or usefulness of various risk-informed projects.

VSL is a way of making a comparison of different attributes or concerns related to the risk issue, but it does not encompass the full set of considerations, including ethical dimensions, associated with life, quality of life, and well-being. For example, the use of VSL has been criticized for how the associated computations consider dimensions of demographics and socioeconomic factors, thereby raising concerns about fairness and equity/equality. The concept of VSL itself has also been criticized in reference to alignment with particular cultural and societal values.

The insurance industry is also known for applying similar methods when evaluating risk. These methods typically estimate the potential financial impact of an

 DOI: 10.1201/9781003437031-12

individual's death or other health consequences. The actuarial calculations may leverage statistical models and historical data to quantify the likelihood of various consequences.

At the operational level, industries may evaluate environmental risk based on the potential financial impact of environmental events, such as natural disasters, pollution, climate change, and ecosystem degradation. They may analyze available data and models to quantify the likelihood and severity of environmental events that can lead to insurance claims. The consequences could be measured by considering property damage, business interruptions, liability claims, and cleanup costs.

It is common for insurance and investment analyses to mostly focus on some financial incentive. While risk events can be insured and corporate initiatives can be backed by investor sentiment, these generally cover only financial aspects. When risk events happen within health, safety, and the environment, often, no program or initiative can fully replace or undo these types of damage.

At the corporate level, many competing objectives are related to a business activity. For example, consider the exploration, drilling, and production of oil and gas operations in offshore drilling operations, as discussed in the case study below.

BP OIL SPILL – DIMENSIONS OF RISK FOR OIL AND GAS OPERATIONS AND RELATED REGULATION

It is likely that BP, like many other companies, conducted internal risk assessments and cost–benefit analyses as part of its decision-making process prior to the 2010 oil spill. While many details of BP's internal decision-making process are not publicly available, we can leverage available reports and general industry understanding to explore the balance between costs and benefits related to the oil spill.

First, consider the cost considerations for a similar company conducting its own risk assessments. An oil and gas company likely evaluates its drilling operations' financial costs and benefits. This includes assessing the projected revenue from oil production, operational expenses, exploration and drilling costs, and potential profits. Concurrently, there is a need to consider the safety of employees, wildlife, and other stakeholders. These assessments aim to balance the costs of implementing safety measures with the potential consequences of accidents. Risk assessments for oil drilling would likely consider the costs associated with compliance with environmental, health, and safety regulations and any penalties or fines for noncompliance.

These types of operations may consider the environmental impacts of their operations, such as those related to the skimming operations as shown in Figure 12.1. This includes assessing risks to ecosystems, wildlife, and protected areas. Environmental impact assessments help identify mitigation measures, including best practices for preventing spills or minimizing their environmental impact.

The National Commission on the BP Deepwater Horizon Oil Spill and Offshore Drilling report (2011), used to support details in this case, provides insight into what type of risk management regime existed before the oil spill and what aspects of their risk activities and operations led to the oil spill. What is known is that BP had some risk management regimes in place before the spill.

FIGURE 12.1   Gulf of Mexico skimming operations near the Deepwater Horizon incident on May 22, 2010. National Archives (2010) photo no. 7854366.

The BP report describes high uncertainties and dangers associated with deepwater drilling:

> But drilling in deepwater brings new risks, not yet completely addressed by the reviews of where it is safe to drill, what could go wrong, and how to respond if something does go awry. The drilling rigs themselves bristle with potentially dangerous machinery. The deepwater environment is cold, dark, distant, and under high pressures – and the oil and gas reservoirs, when found, exist at even higher pressures (thousands of pounds per square inch), compounding the risks if a well gets out of control. The Deepwater Horizon and Macondo well vividly illustrated all of those very real risks. When a failure happens at such depths, regaining control is a formidable engineering challenge – and the costs of failure, we now know, can be catastrophically high.

These uncertainties manifested themselves in many high-profile oil- and gas-related incidents. For example, in 1980, the *Alexander Kielland* drilling rig in the Norwegian North Sea capsized, resulting in 123 deaths. This and several related events were attributed to poor safety controls, structural failures, insufficient worker training, and weak emergency management/communication measures. As a result of these incidents, the Norwegian government shifted the regulatory perspective to requiring operators to demonstrate risk assessment, safety, and appropriate risk management initiatives. This was in contrast to a prescriptive regulatory approach, such as in the United States, which required industry to adhere to regulations and inspections despite the newness of and high uncertainties with drilling activities.

The BP report further simplifies the causal factors related to the well blowout:

> The immediate cause of the Macondo blowout was a failure to contain hydro-carbon pressures in the well. Three things could have contained those pressures: the cement at the bottom of the well, the mud in the well and in the riser, and the blowout preventer. But mistakes and failures to appreciate risk compromised each of those potential barriers, steadily depriving the rig crew of safeguards until the blowout was inevitable and, at the very end, uncontrollable.

The report describes cementing the oil well as an "inherently uncertain process." Because the crew had limited visibility of the cement, as it worked far underwater, there was limited understanding of whether the cementing job was safely and correctly completed. There was also insufficient experience leveraging a nitrogen foam cement technology used in the operation. There were also apparent uncertainties about the design specifics of the cement column, as BP's internal guidelines differed from the relevant regulations.

The report also describes deficiencies in the crew, specifically the inability to recognize the severity of the risk event and the failure to react appropriately. The crew also had limited time to act and had inadequate emergency response training. Due to last-minute changes to the design, typical checks and peer reviews did not sufficiently evaluate risks related to the implemented design and procedures. A severe lack of communication within BP and across its partners is also considered to be a causal factor. Many of the decisions that contributed to the risk event may have been motivated by saving time and/or money.

BP's Oil Spill Response Plan, mandated by the Oil Pollution Act of 1990, was accused of being poorly conducted and cursory. Segments of the report were allegedly copied from government websites and contained clear inaccuracies resulting from the lack of attention. Government agencies were also criticized for insufficient review of the drilling operations, such as in relation to the Clean Water Act.

Following the oil spill, BP faced several significant repercussions. The repercussions included significant costs for the immediate response to the spill and subsequent cleanup efforts. In addition, the environmental damage was extensive, impacting ecosystems, wildlife, and natural resources. There was extensive loss of habitats, impacts on marine life and fisheries, and long-term ecological effects. The valuation of environmental damages is a complex task that requires considering factors such as ecological services, biodiversity loss, and habitat restoration costs, which are not easily monetized. The oil spill also led to economic losses in fishing, tourism, and coastal businesses. The legal liabilities and penalties for the oil spill were also extensive, including fines under the U.S. Clean Water Act and legal settlements with various parties. It's unclear whether the extent of these costs was foreseen before the risk event.

The Deepwater Horizon oil spill led to increased scrutiny of risk assessment, risk management, and safety practices in deepwater drilling operations. The spill prompted an evaluation of safety regulations and practices in offshore drilling. Regulatory bodies have implemented stricter safety standards and more rigorous oversight of drilling operations. This includes requirements for blowout preventers, well design and construction, and emergency response preparedness. There is also an increased emphasis on safety culture through employee training and reporting of safety concerns, as well as in safety-related technologies related to blowout preventers and well-monitoring systems.

This case study highlights that quantified costs and benefits are a single component of a context-based holistic perspective on an organization's risk assessment and management regimes. Putting monetary amounts on aspects like biodiversity and safety is one way of evaluating costs. However, it ignores ethical, reputational, and other aspects of the risk situation, which are also very real and very value-based.

This case study also highlights that regulation is slow and often reactionary to risk events. The increased safeguards introduced through new regulations only had sufficient interest and political backing after the BP oil spill had occurred. This is something to consider for any new or existing activity involving significant uncertainties and extreme threats to life and the environment.

The issues presented in the BP case study suggest that there is a need for continued emphasis on factors related to risk that extend far beyond monetized units.

In recent years, there has been increased attention on developing holistic perspectives on a firm's stances toward risk issues. While there are many competing initiatives, such as socially responsible investing or green investing, the most well-known is Environmental, Social, and Governance (ESG). ESG refers to an evaluation of a company's sustainability and ethical practices. The environmental component considers factors such as the company's carbon footprint, emissions, impact on biodiversity, and other metrics related to environmental health. The social component considers relationships with various stakeholders, such as labor practices and outreach to communities. The governance component considers factors such as board composition and corporate ethics. Many of the risk case studies presented in this book encompass these issues, such as the Flint sit-down strike, the Dhaka garment factory fire, and the BP oil spill.

While the intent of these initiatives is likely to pursue long-term goals that aid in understanding and managing risk in coordination with other goals, these initiatives are not a substitute for carefully implemented risk assessment and management regimes. There are no one-size-fits-all approaches for these types of risk applications, and trusting that evaluation to an outside evaluator with little firm or industry-specific knowledge is problematic. These initiatives have also been accused of greenwashing, as firms can "check off the boxes" to meet ESG-related criteria without creating meaningful risk-related programs and using risk principles to make decisions in their firm. Factors considered in ESG initiatives can also be somewhat ambiguous or, conversely, quite complex, making it difficult to make meaningful measurements of abilities to meet various criteria.

The conversation about these various dimensions of risk serves as a reminder that one can check off boxes showing compliance with various policies and standards. However, what matters is that the risk science is of high quality. Over-focusing on meeting particular criteria can undermine more extensive risk-related efforts. High-quality risk science can instead be used to check off those boxes, but a holistic look at the firm and its operations with a broader context is needed in the long run.

WORKS CITED

National Commission on the BP Deepwater Horizon Oil Spill, & Offshore Dril (Eds.). (2011). Deep water: The gulf oil disaster and the future of offshore drilling: Report to the President, January 2011: The Gulf Oil Disaster and the Future of Offshore Drilling. Government Printing Office.

National Archives (2010). GULF OF MEXICO, May 22, 2010 – Vessels conduct skimming operations in the Gulf of Mexico near the site of the Deepwater Horizon incident May 16, 2010. The Deepwater Horizon, a mobile offshore drilling unit, exploded April 20, 2010 and sank two days later. U. S. Navy photo by Stephanie Brown, Naval Sea Systems Command. https://catalog.archives.gov/id/7854366

13 The Weight Given to Resilience

The concept of resilience has taken on meaning across many fields and disciplines. The concept generally refers to the ability to withstand and recover from various types of disruptions. Those disruptions may be sudden or gradual and may come from internal to the system or some external source.

Various industries and professions have different definitions and uses for resilience-related terms. For example, in an environmental setting, resilience refers to the capacity of ecosystems to survive and function when subject to risk events like natural disasters, climate change, or human activities. Similarly, in engineering settings, resilience refers to the ability of infrastructure to withstand and recover from major risk events. Social applications use slightly different definitions, such as those related to community abilities to recover from social, political, and economic disruptions or, at the individual level, the ability to recover from various traumas.

One can invest in system resilience in ways that improve overall system resilience across various disruptive events. In other words, resilience investments can be effective in settings even if a particular event was not foreseen.

As this chapter will demonstrate, resilience is often developed over time. Experience with past risk events and knowledge of the system that develops with time, combined with innovative practices and technologies, are often the mechanisms that build resilience.

Businesses often invest in resilience activities for resilience purposes or general operational wellness. For example, it is common to conduct business continuity planning to develop protocols to maintain essential operations during disruptions or large-scale emergencies. Similarly, crisis management and response plans can help manage significant risk events.

Resilience is particularly important for securing critical infrastructures. Critical infrastructure sectors include critical manufacturing, defense industrial bases, emergency services, energy, communications, food/agriculture, healthcare, water, and transportation systems. Infrastructure resilience investments are often highly visible, including efforts for upgrading and retrofitting infrastructures, implementation of smart technologies, installation of supplemental power systems, design adaptations for climate change (e.g., sea level rise, increased temperatures, and extreme weather events), enhanced emergency response systems, and public awareness campaigns.

However, resilience does not necessarily develop independently without meaningful investments, managerial attention, and stakeholder will. Hurricane Katrina provides an example of resilience in the aftermath of a natural disaster that severely impacted a major city. The following case study describes the rebuilding process and how the event prompted additional insights and initiatives for disaster preparedness, eventually promoting improved abilities to withstand and recover from disasters.

 DOI: 10.1201/9781003437031-13

HURRICANE KATRINA – WHY NEW ORLEANS WAS VULNERABLE AND HOW RESILIENCE EMERGED

The August 2005 Hurricane Katrina was one of the most fatal and most costly hurricanes in U.S. history, striking the United States Gulf Coast. Katrina made landfall as a Category 3 storm crossing over southeastern Louisiana and had a devastating impact across the region, including the city of New Orleans. Some of the area is located below sea level and is protected by an extensive system of levees and canals. The city's vulnerabilities due to location, aging infrastructure, and insufficient maintenance made it particularly susceptible to flooding and devastation.

The storm caused the failure of several levees and floodwalls in the New Orleans area. This led to catastrophic flooding, submerging a significant proportion of the city and leaving thousands of residents stranded on rooftops or in shelters, such as the shelter shown in Figure 13.1, and the damage is shown in Figures 13.2 and 13.3. The storm caused somewhere between 1,200 and 1,800 deaths, and over one million people were displaced. The storm caused extensive damage to homes, businesses, and critical infrastructure, including schools, hospitals, and transportation systems, leading to a total economic cost of about $125 billion. In addition, the storm had severe environmental impacts, including oil spills, contamination of water sources, and damage to ecosystems. The impact of Hurricane Katrina disproportionately affected marginalized communities where residents lacked resources and had limited access to transportation.

The evacuation process in preparation for the storm faced significant challenges because many residents did not have access to transportation or could not leave due to personal reasons. In addition, the response and coordination of local, state, and

FIGURE 13.1   Hurricane Katrina survivors from New Orleans at an emergency shelter in the Houston Astrodome. Wikimedia Commons (2005). Andrea Booher/FEMA. Orginal Source: FEMA Photo Library.

FIGURE 13.2 Flooding in New Orleans following Hurricane Katrina, August 29, 2005. National Archives (2005) photo no. 5694700.

federal agencies in the aftermath of Hurricane Katrina were widely criticized. As a result, the storm highlights the need for more attention toward disaster response strategies and planning, improvements in infrastructure protection and maintenance, and the establishment of more robust warning systems and community engagement.

The emergency shelters were said to be disorganized and did not have sufficient resources to operate. Law enforcement agencies struggled to keep order as violence and looting were allegedly prevalent.

The Katrina disaster prompted cities and nations across the world to carefully consider established disaster response and preparedness protocols. There was intense discussion about the need for improved infrastructure and resources to mitigate the impact of future hurricanes. One example of the resilience efforts includes a comprehensive overview of the city's flood protection infrastructure. Over the decades since the disaster, there has been a focus on strengthening and improving levees, floodwalls, and pumping stations to reduce the risk associated with future flooding. In addition, there has been increased attention on the city's evacuation plans, including better coordination among agencies, improved management of resources, and more effective communication strategies. During the rebuilding process, there was an emphasis on reevaluation of urban planning in New Orleans, including a focus on smart growth, sustainable design practices, green infrastructure, and other initiatives.

After Hurricane Katrina, New Orleans experienced a complex process of gentrification. Some long-term residents, particularly those who faced challenges in rebuilding their homes, were unable to return and were forced to relocate to other areas. The gentrification exacerbated existing disparities in the recovery process.

FIGURE 13.3 Comparison of before vs. after Hurricane Katrina on U.S. Highway 90. National Archives (2005) photo no. 5695518. From National Archives (2005). Photographed by FEMA/Mark Wolfe. https://catalog.archives.gov/id/5695518

One also cannot view Hurricane Katrina in isolation from other post-9/11 initiatives to improve abilities to prevent, protect, mitigate, respond, and recover from risk events in general. Shortly before Hurricane Katrina, the National Incident Management System (NIMS) was established in the United States. The NIMS further promotes a standardized approach to coordinate incident management. The plan calls for coordinated communication at all levels of emergency response, including strategic decision-making, on-site response, and public outreach. Similarly, the Emergency Response Coordination Center (ERCC) in the European Union coordinates assistance in disaster situations, covering many scenarios including hurricanes, wars, floods, and oil spills.

It is unclear whether the consequences of Hurricane Katrina and the subsequent regulation have improved resilience practices. In 2012, after the rebuilding was complete, Hurricane Isaac struck the Gulf Coast, including New Orleans. Some say the lessons learned from Hurricane Katrina significantly influenced the response to Isaac. In preparation for Hurricane Isaac, evacuation plans and communication protocols were said to be stronger. There was said to be more clear emergency preparedness guidance and better transportation options for residents who needed to evacuate. The city's flood protection system was also improved, as the upgraded levees helped mitigate the impact of storm surges and flooding. In addition, local, state, and federal agencies collaborated to coordinate response efforts. There were also notable improvements in communication and information dissemination to the public during the hurricane.

Outside of New Orleans, there are also signs of improved (but still in need of further improvement) hurricane preparedness and response activities. For example, consider the response to Hurricane Maria, which struck Puerto Rico and other parts of the Caribbean in September 2017. Despite the disastrous outcome, the federal government's response to Hurricane Maria is said to have been faster and more effective than its response to Hurricane Katrina. In addition, the Internet and social media enabled real-time communication and assistance efforts. While, in general, the response to Hurricane Maria was more coordinated compared to that of Hurricane Katrina, there remained several large challenges. Similar to the criticism of Hurricane Katrina, there were delays in delivering aid and resources to residents. Many residents reported being without essential supplies, such as food, water, and medical assistance, for an extended period. There were also communication and coordination challenges among federal, state, and local agencies, causing delays in distributing aid and support to affected communities. Puerto Rico's already vulnerable power and infrastructure systems were severely damaged by Hurricane Maria, leading to widespread power outages and difficulties in delivering aid to remote areas.

In addition, modern technologies can promote resilience in ways that were impossible even a few years ago. For example, technological enhancements including sensors, satellite imagery, and drones help monitor conditions in preparation for, during, and in response to risk events. Social media platforms also enable real-time information sharing during disasters in ways that are crowdsourced from residents. In the future, many promising new technologies can have an even greater impact on disaster recovery and resilience operations. For example, AI algorithms can analyze large datasets to predict the impacts of disasters and optimize resource distribution. In addition, drones equipped with cameras and sensors can assess disaster-affected areas, locate survivors, and deliver essential supplies to remote or inaccessible locations.

This chapter has discussed the importance of resilience when considering risk. This resilience can apply to natural disasters, but similar concepts also apply across various scenarios, such as war, terrorism, and climate change. Thus, resilience is a necessary component of a complete risk science process. The following chapter will consider two additional case studies that help piece together the various components of the themes of this book to actual risk events.

WORKS CITED

Wikimedia Commons (2005). Andrea Booher/FEMA. Original Source: FEMA Photo Library. https://commons.wikimedia.org/wiki/File:Katrina-14451.jpg

National Archives (2005). [Hurricane Katrina] New Orleans, LA, August 29, 2005 – The breach in the 17th Street canal levee causing flooding in the city following Hurricane Katrina. Photographed at 6:43 PM. Marty Bahamonde/FEMA https://catalog.archives.gov/id/5694700

National Archives (2005). [Hurricane Katrina] Biloxi, Miss., September 3 and November 3, 2005 – U.S. Highway 90 before (top) and after repair from damage caused by Hurricane Katrina. FEMA/Mark Wolfe. https://catalog.archives.gov/id/5695518

FURTHER READING

Aven, T. (2019). The call for a shift from risk to resilience: What does it mean? *Risk Analysis*, 39(6), 1196–1203.

FEMA. (2018). *Hurricanes Irma and Maria in Puerto Rico: Building Performance Observations, Recommendations, and Technical Guidance*. FEMA. https://www.fema.gov/sites/default/files/2020-07/mat-report_hurricane-irma-maria-puerto-rico_2.pdf

Hollnagel, E., Woods, D. and Leveson, N. (2006). *Resilience Engineering: Concepts and Precepts*. London: Ashgate.

United States. Congress. Senate. Committee on Homeland Security, & Governmental Affairs. (2006). Hurricane Katrina: A nation still unprepared: Special Report of the Committee on Homeland Security and Governmental Affairs, United States Senate, Together with Additional Views (Vol. 109, No. 322). United States Senate.

14 The Big Picture of Risk Science Surrounding Major Historical Events

This chapter will relate the concepts of this book to major risk issues and events.

First, we consider tornadoes, whose deadly impacts have marked history. Understanding of tornadoes has primarily developed over time with scientific and technological progress. Tornadoes are illustrative of the big picture of risk science because it is possible to analyze the evolution of risk assessment and management surrounding tornadoes in parallel with the evolution of scientific knowledge about these events.

The topic of tornadoes is a case example of risk science in action. These events involve large amounts of dread due to their catastrophic potential, leading to continued awareness and emphasis on preparedness. In the past, tornadoes were characterized by large uncertainties and low knowledge. However, time has shown increased knowledge about these phenomena and improving measures to mitigate the risk.

TORNADOES – THE BIG PICTURE AROUND RISK ISSUES

Tornadoes have had a long-documented history of destruction. For example, consider the Tri-State Tornado of 1925, the deadliest tornado in U.S. history. This tornado struck parts of Missouri, Illinois, and Indiana with a path of approximately 219 miles, causing over 695 fatalities. As a more recent example, consider the Joplin tornado, an EF5-rated tornado that struck Joplin, Missouri in 2011. This tornado had a death toll of 161 people and an estimated property damage of about $2.8 billion. While tornadoes are most common in the southern United States, they have been documented across the world.

The understanding of tornadoes has evolved through scientific research, observations, and meteorological advancements. In the late 1800s, scientist John P. Finley developed a classification system for tornadoes and developed a set of formal rules for predicting tornadoes. While Finley's progress in tornado research was undermined by political and bureaucratic factors, there was a continued need for more understanding of the phenomenon. In the mid-1900s, radar technology was developed, allowing more careful data collection. In the 1970s, the Fujita scale was introduced to quantify tornado intensity further. Later, Doppler radar, which measures the velocity of moving particles within storms, revolutionized tornado detection and monitoring, allowing for increased accuracy in predicting potential for tornadoes. The development of supercomputers and advanced numerical models have also allowed scientists to understand tornadoes further, such as by using models to

 DOI: 10.1201/9781003437031-14

simulate tornado formation and study the observed phenomena. As a result, recent years have drastically improved tornado detection and preparedness. For example, over the past decades, the National Weather Service (NWS) in the United States developed a severe weather warning system, including tornado watches (indicating conditions favorable for tornadoes) and tornado warnings (indicating an imminent threat).

Tornadoes are notable because we can learn from them how these risk events have been viewed over time as knowledge grows and as technologies develop. When understandings of tornadoes were characterized by relatively low knowledge and high uncertainty, there was an extensive history of misinformation and disinformation used to understand them. For example, a pioneer in understanding tornadoes, J.P. Finley, was caught in political turmoil over his research (Galway, 1985). The referenced work states:

> The combination of Finley's apparent success at predicting tornadoes and his persistent belief that these predictions should be incorporated into official releases (indications) of the Signal Corps, elicited instructions issued by the chief signal officer in 1885: When the current weather report favored the occurrence of tornadoes, the indications would contain a special warning that violent local storms were indicated for the area of concern. The instructions also stated that the word tornadoes would not be used.

A chief signal officer stated, "it is believed that the harm done by such a prediction would eventually be greater than that which results from the tornado itself."

These quotes suggest that accuracy posed a greater threat than potential inaccuracies and falsehoods associated with a storm in terms of its potential to induce fear and panic. One can ask if this evidence suggests that misinformation or lack of information were preferred over the development and reliance on accurate predictive models. Perhaps avoiding discussion of tornado prediction also avoided inciting fear and panic.

There may also have been concern over promoting scientific knowledge and explainable phenomena as the contributors to tornadoes, as opposed to relying upon existing narratives. Generally, in phenomena that are poorly understood from a scientific perspective, there may be a reluctance to accept new scientific reasoning, particularly when those reasonings conflict with preexisting beliefs and narratives. Also in the case of J.P. Finley, there was a question of political/bureaucratic influences, and generally who should be responsible for creating narratives around those scientific perspectives. In some respects, one can say that misinformation and disinformation can be rampant and potentially go unnoticed when there is a lack of a consensus and strong knowledge base for a risk topic. However, as knowledge grew, so did capacities to make meaningful and specific improvements to practices, such as building design and preparedness.

Consider how tornadoes were viewed during the Cold War era as an example of misinformation and disinformation. There was speculation that tornadoes resulted from nuclear testing. This was a time when nuclear testing was in its early stages, and nuclear weapons and their potential effects were generally poorly understood.

When tornadoes coincidentally occurred near the locations and aligned with time frames of atomic testing, people may have erroneously linked the two events. However, tornadoes have been occurring since long before the development of atomic weapons, making it unlikely that the tornado activity was related to nuclear testing. This misinformation was potentially rooted in fear, conspiracy theories, and the geopolitical climate of the time. In addition, media coverage, rumors, and other narratives about nuclear weapons and their impacts on the environment may have contributed to the misinformation. While there is little evidence of the misinformation being used for political purposes or for the gain of some parties, it's possible that the resulting fears and controversies could be exploited for some parties' gain.

With an increased understanding of tornadoes, there is also improvement in addressing the risk associated with tornadoes. Over the past many decades, science, engineering, and policy have worked to develop codes and practices that can improve abilities to withstand tornadoes. One major risk mitigation measure has included building codes that intend to increase the safety of occupants during severe weather events, not only limited to tornadoes. For example, practices in tornado-prone areas include materials and design features that can withstand high winds and flying debris, such as reinforced roofs, impact-resistant windows and doors, safe rooms, and storm shelters. In addition, risk mitigation measures sometimes involve installing tornado warning systems in commercial and public buildings to alert occupants of impending tornadoes and allow them to take appropriate safety measures. These initiatives are also increasingly coordinated with other measures for tornado preparedness, such as education for residents and building occupants. This may include tornado safety drills, the development of tornado safety plans, and the dedication of shelter locations.

In the case of the Joplin tornado, the warning occurred 17 minutes before the tornado struck (see Figure 14.1 for imagery). This allowed time for residents to seek shelter. However, as a NIST report (Kuligowski et al., 2014) uncovered, there was general confusion over the interpretation of the siren and what constituted an all-clear signal at the end of the warning. Residents were also desensitized to sirens due to frequent testing and may have been confused over the distinction between testing and an actual emergency. There were concerns over the general understanding of the threat communicated with the siren.

The Joplin tornado was an impetus for improvements in disaster preparedness and mitigation. The destruction caused by the tornado provided an opportunity to rebuild with a focus on resilience, safety, and community needs. For example, rebuilding homes, businesses, and public buildings emphasized tornado-resistant construction techniques and materials, including stronger foundations, reinforced walls, and improved roofing systems. There were also upgrades to utilities and services. Electrical and water systems were modernized, with more robust infrastructure implemented to increase reliability and resilience. Efforts were made to enhance communication systems, including emergency alert systems, to ensure effective communication during severe weather events.

The Joplin tornado also prompted improved abilities to mitigate, prepare, respond, and recover from various risk events, not only tornadoes. The tornado prompted a review of emergency management and disaster preparedness practices. For example,

FIGURE 14.1 Debris removal in Joplin after an EF5 tornado. National Archives (2011) photo no. 7858012.

the National Institute of Standards and Technology (NIST) report emphasizes the importance of warning systems that incorporate more current technologies, such as initiatives for pushing alerts to mobile phones and social media using geographic locations, which can reduce reliance on sirens. More generally, work is ongoing to update building codes and ensure proper shelter space in existing buildings. These initiatives and lessons learned from past disasters promote improvements in emergency response protocols, community resilience planning, and coordination among local, state, and federal agencies.

However, residential and structural preparedness is not to be seen in isolation from larger factors. For example, consider the occupational safety-related implications in cases where tornadoes strike workplaces. There are several historical examples of tornadoes impacting warehouse facilities in which workers were unaware of emergency protocol warning systems and did not have access to cellular communication devices.

There are also concerns over popular and cost-effective design practices for warehouses and retail establishments. For example, the popular "tilt-up" design has been criticized for being unable to withstand tornadoes and earthquakes appropriately. If a roof is lost, possibly due to high winds, the walls lack sufficient stability, creating conditions for collapse. As a result, some weather-related risk communications discourage sheltering in warehouses and auditoriums with wide-span roofs.

These past tornado events have also spurred conversations about how employers should prepare and respond to emergency warnings. Also, there appear to be inconsistencies in requirements for storm shelters in warehouses. With increased concern about natural disasters, there remains the question of whether the existing measures for disaster mitigation, preparedness, response, and recovery are sufficient for the future.

While communities and policymakers have improved their abilities to understand and respond to those tornadoes, there remains more work to be done.

The second major case involves three global health priorities.

SMOKING, SOCIAL MEDIA, AND PROCESSED FOODS – LINKING THREE GLOBAL RISK ISSUES

There has long been discussion about the cigarette industry and the global impacts of smoking. In this section, we consider that history in parallel with the use of technologies, mainly social media use, and more recent concerns over processed foods. The parallels can provide some interesting perspectives on risk science associated with basic behaviors. When these behaviors have detrimental health outcomes, there is a question about how individuals and societies should address risk while considering the optimal balance with regulation and widespread governance of those behaviors. Thus, this balance can determine whether the risk is addressed versus whether mitigation measures undermine other essential elements of business and society.

The cultivation of tobacco is documented across several centuries. During the Industrial Revolution, the tobacco industry leveraged mass production of cigarettes by using machinery for rolling and packaging them. Soon after that, multinational tobacco corporations were formed.

In the mid-1900s, smoking was often portrayed in movies, television shows, and advertisements. In addition, famous celebrities were often used in tobacco advertising campaigns, further endorsing the glamorization of cigarettes, as shown in Figure 14.2. Smoking was prevalent in social settings, such as bars, vehicles, and airplanes. In addition, marketing strategies employed witty slogans, attractive packaging, and sometimes, bold (now clearly debunked) claims about cigarette use and health.

Cigarette use was also leveraged to make statements about major societal issues. For example, smoking was advertised and often seen as an act of rebellion against authority. Even governments promoted the use of smoking. For example, during World War II, cigarettes were often included in soldiers' rations. Cigarette sales also generated revenue through taxes.

Over time, scientific evidence developed to provide a more clear association between cigarette use and health impacts. In particular, there was substantial evidence of the harmful effects of smoking on health, including the risks of cancer, heart disease, and respiratory ailments.

While this text does not summarize the legal details, scientific research, manipulation of scientific research, and other controversies, one can generalize and say that several lawsuits have shed light on a disassociation between scientific knowledge about cigarette use and the knowledge shared with individuals consuming the products. These health-related linkages have been developing since the 1950s. By the 1960s, the Surgeon General's report in the United States officially acknowledged the health risks associated with smoking.

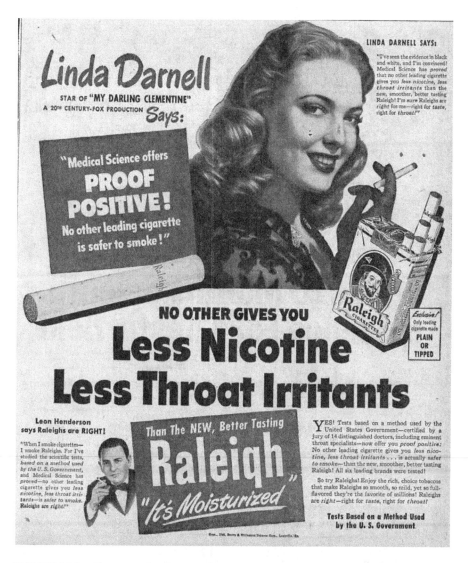

FIGURE 14.2 Cigarette advertisement from 1949. Stanford University - Stanford Research Into the Impact of Tobacco Advertising (1949).

Nicotine addiction was one factor behind smoking behaviors. Also, many tobacco ads, such as those available through Stanford research (2024), promoted the idea that smoking was an individual choice, suggesting that responsibilities associated with health consequences rest with the individual choosing to use the product, relating heavily to the concept of voluntariness discussed in Chapter 10. Similarly, some ads conceptualized the protection of freedom to smoke as a parallel to other freedoms. As shown in Figure 14.3, smoking ads also were a venue for scientific health claims.

Various nations have addressed cigarette risks through regulatory measures and public health campaigns. Most commonly, there has been widespread use of tobacco taxation, smoke-free laws, graphic health warnings on products, and restrictions on

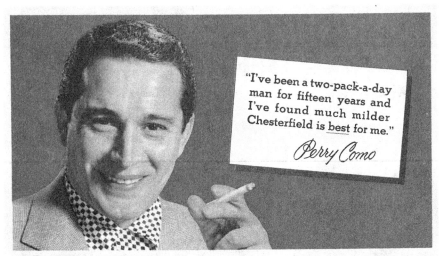

FIGURE 14.3 Cigarette advertisement from 1953. Stanford University - Stanford Research Into the Impact of Tobacco Advertising (1953).

advertising. Another effective tool to address misinformation and disinformation about smoking has been public health campaigns that raise awareness about the health risks of smoking.

Next, consider social media. While social media has been found to have several positive effects, such as providing a medium for social support, education, and communication, there are also concerns related to potentially negative implications of social media. Individuals may use social media to seek social validation through various types of interactions. Some hypothesize that individuals may seek out notifications (e.g., "likes" and comments), receiving cognitive rewards for those activities. As a result, individuals continuously anticipate additional rewards; they are incentivized to check for additional notifications continuously.

Additionally, these platforms can be used for purposes of social comparison in which users compare themselves to others, prompting users to stay active on the platform. Similarly, there can be a tendency to elevate controversies, potentially promoting trolling, debates, and arguments that keep users engaged. Some platforms and web resources use infinite scrolling, as there is no end to additional or new content during the scrolling process, promoting prolonged engagement. These platforms may also use algorithms to align content with the user's engagement history, ensuring that users continue scrolling.

This nature of social media poses several concerns related to health and well-being. For example, excessive use of social media can take time away from other meaningful activities. It can also disrupt sleep. Cyberbullying remains a consistent issue, which can lead to serious health outcomes for victims. Social media can also promote body image issues, as the use of photo filters and photos of other individuals can contribute to body dissatisfaction and other mental and behavioral concerns. As discussed widely throughout this book, social media can also be a source of misinformation and disinformation for major risk topic areas.

As a third example, consider ultra-processed foods. These foods are sometimes engineered to trigger sensations of pleasure and reward. Repeated consumption of those foods can lead to a desire for more, creating a cycle of consumption. These foods leverage additives (such as those described in previous chapters of this book) and textures to make foods more appealing. The consumption of these foods can potentially have severe health impacts.

Many of these foods are widely accessible and marketed, including to children. For example, consider food advertising aimed at children. In addition, consider how certain foods are shown and consumed in various forms of media (e.g., TV, movies, and social media). These foods can also be intentionally placed at eye level (kids' or adults') in grocery stores in order to promote sales.

As concerns grew about the impact of ultra-processed food on children, there were initiatives to introduce self-regulatory measures or restrictions on ultra-processed foods and related advertising. For example, the World Health Organization and other health organizations have called for stricter regulations on junk food advertising to kids to protect health and to promote healthier eating habits. More recently, nations worldwide have considered bans on advertising processed foods, particularly to children.

There are several key connections among these three very different risk areas: smoking, social media, and processed foods.

These examples illustrate that knowledge of health outcomes from these risk-related activities is critical for how various stakeholders understand and manage the risk; and that knowledge can vary based on many factors. This is a strong concern from a risk perspective, as risk science relies on appropriately disseminating key knowledge. The suppression or manipulation of knowledge can exacerbate risk in ways that can have severe consequences regarding the ability to manage risk.

In addition, even when knowledge becomes more balanced, there is recognition that risk communication does not necessarily result in a policy change or modifications to individual behaviors. However, as discussed in this book, accepting the risk is also a form of managing risk.

We can also further relate these issues to concepts of voluntariness and controllability, as discussed in Chapter 10. The perception of risk related to the activities discussed in this chapter may be lowered due to the voluntary nature of the activities (e.g., whether to engage in social media and whether to consume certain foods). In addition, individuals may feel a sense of control over their exposure to the risk (e.g., by choosing particular ways to use social media).

At a larger level, these examples suggest distinct challenges in identifying the role of individuals in managing these risk issues. Whether regulation should impact the behaviors of others, whether coercion of individuals to undertake risk activities is legal or ethical, and whether disparities in knowledge across parties are an acceptable approach remain areas for inquiry and debate.

This chapter has presented case studies incorporating key topics from this book. The concept of tornadoes provides a clearer picture of risk science in action, particularly as abilities to understand and treat the risk improved over time. The concepts of smoking, social media, and ultra-processed foods, however, raise many questions in the realm of ethics for which risk science offers limited guidance. The lack of guidance on these ethical concerns is large because risk science aims for transparency and responsibility in decision-making but does not prescribe the actual decisions. The following chapters will elaborate more on the value of risk science and areas that are largely out of the scope of risk science.

WORKS CITED

Galway, J. G. (1985). JP Finley: The first severe storms forecaster. *Bulletin of the American Meteorological Society*, 66(11), 1389–1395.

Kuligowski, E. D., Lombardo, F. T., Phan, L., Levitan, M. L. and Jorgensen, D. P. (2014). Final report, National Institute of Standards and Technology (NIST) technical investigation of the May 22, 2011, tornado in Joplin, Missouri. https://nvlpubs.nist.gov/nistpubs/NCSTAR/NIST.NCSTAR.3.pdf

National Archives (2011). Severe Storm^Tornado – Joplin, Mo., August 7, 2011 – Expedited debris removal is completed in Joplin after an EF5 tornado struck the town. FEMA is in Joplin to provide assistance to disaster survivors. Suzanne Everson/FEMA. https://catalog.archives.gov/id/7858012

Stanford. (2024). The Stanford research into the impact of tobacco advertising (SRITA) collection. https://tobacco.stanford.edu/

Stanford University – Stanford Research into the Impact of Tobacco Advertising (1949). Less Nicotine – img3201. https://tobacco.stanford.edu/cigarette/img3201/

Stanford University – Stanford Research into the Impact of Tobacco Advertising (1953). Pseudoscience – img1552. https://tobacco.stanford.edu/cigarette/img1552/

FURTHER READING

Office of the Surgeon General. (2021). Protecting youth mental health: The US surgeon general's advisory [Internet]. https://www.hhs.gov/sites/default/files/surgeon-general-youth-mental-health-advisory.pdf

Peterson, A. (2024). The new science on what ultra-processed food does to your brain. https://www.wsj.com/health/wellness/ultra-processed-food-brain-health-7a3f9827

15 How Risk Science Improves the Ability to Address the Gaps in the Themes Presented

The concepts and practice of risk science have drastically improved over time. New innovations and insights have often resulted from learning by experience. For example, histories of understanding and witnessing tornadoes have contributed to ever-increasing abilities to mitigate risk related to tornadoes.

One of the most common misconceptions about risk science in general is the focus on the ability to predict the future. As there is no crystal ball, one cannot predict all future events. However, modern scientific knowledge has made great progress in prediction in some cases, such as related to the weather. Sometimes, lack of sufficient accuracy is seen as a deterrent to addressing risk in general. While predictions are often incorporated into the practice of risk science, those predictions are accompanied by other factors. For example, there is concurrent consideration of uncertainties, as one can recognize that any future prediction may not match actual events. These predictions often involve mental or quantitative models, and those models have intrinsic errors and are only simplifications of a complex reality. In addition, there is concurrent consideration of knowledge, as some predictions can be founded on detailed scientific understanding, while others can have little basis. Knowledge strength is an important factor to consider when making risk-based decisions.

Because knowledge is a basis for risk characterization and decision-making, the issue of poor integrity of the evidence informing that knowledge is problematic. We recognize that one can have a high-quality risk study even if the evidence is poor, but only if those evidence issues are communicated and made transparent to decision-makers. Instead, the historic risk events studied in this book have shown that poor integrity of evidence can be concealed, leading to knowledge appearing misleadingly strong, when in fact, it is not. History provides examples of this phenomenon. For example, consider the associations between nuclear testing and tornadoes, in which some believed there was a causal relationship between the two. Similarly, consider the health claims made for smoking that appeared very confident, but, in reality, were harmful to societal health. These types of claims that appear of high-quality, but are not, can drive decision-making in ways that can be counterproductive and potentially catastrophic.

On a similar note, misinformation and disinformation were prevalent across the historic risk events studied in this book. Misinformation, while an innocent falsehood, can be as dangerous as disinformation. For example, in the COVID-19 case

DOI: 10.1201/9781003437031-15

study, there were cases of dangerous treatments, such as those related to ingesting household chemicals, being touted as cures for the disease. Those treatments may have been equally or more deadly than the disease itself for some individuals. Yet those dangerous treatments gained traction on social media and were shared by individuals in attempts to protect others. Thus, it can be imperative in risk settings to identify misinformation and disinformation and consider obligations to act as a whistleblower when appropriate, but also recognize that there can be limitations in the ability to address or correct misinformation/disinformation when the falsehoods are identified. Equally so, there is a vast gray area in distinguishing the boundaries between information, misinformation, and disinformation, particularly when there are large uncertainties and a poor knowledge base.

The historic risk events studied in this book also showed that in situations with low knowledge and high uncertainties, misinformation and disinformation can quickly spread. Individuals are expected to seek answers and certainty, especially in public-sector situations when those individuals are taxpayers who vote and pay for accountability. When credible sources do not provide answers and certainty, they allow less credible, yet confident, sources to control the narrative. When the narrative is not founded on a high-quality risk study, any attempts at managing the risk can again be undermined.

Risk science also recognizes the power of decision-making. The risk analyst's concern is to characterize the risk, but in the end, risk-based decisions are based on the values of the decision-makers. It is critical for the risk analyst who has a strong understanding of the risk issue to remain separate from the chosen initiatives intending to address the risk. The analyst may personally have come to various conclusions and may be fundamentally attached to particular actions and decisions about how to address the risk. However, the risk analyst's role is to be neutral, intending to aid decision-making in their pursuit of evidence-based decisions.

Risk science recognizes that different people address risk in different ways. That can be expected because all individuals are subject to their own perceptions of risk and differences in values and priorities, based on many factors, including familiarity, dread, media reports, social circles, and more. As a result, those individuals will make or support differing initiatives to address risk.

Risk science also recognizes that the understanding and characterization of risk, combined with decision-making, is not the only component of a comprehensive risk program. It is equally important to have an effective risk culture. Many risk regimes exist. However, none of these replace a risk culture in which all organizational stakeholders feel responsibility for addressing risk. For example, consider the Titanic case study from Chapter 6. Prior to the risk event, a nearby ship sent the message: "heavy pack ice and [a] great number [of] large icebergs," while the Titanic radio operator did not relay the warning. Later, the SS Californian sent a message to the Titanic: "Say, old man, we are stopped and surrounded by ice." The Titanic radio operator responded, "Shut up! Shut up! I am busy. I am working Cape Race." If the radio operator had felt more responsible for the risk, the outcome could have been very different. In many more recent examples, for example, with the Fukushima Daiichi, Deepwater Horizon oil spill, and the East Palestine train derailment cases, risk programs and controls were in place but were not supplemented by sufficient risk culture that supported acting on signs that those programs were not fully aligned with the larger risk-related goals. Most importantly, those existing risk programs also needed to be supplemented by individuals, or whistleblowers, who spoke up when deficiencies were detected.

No risk culture will be perfect from the onset. Risk science relies on the fact that understanding and managing risk will improve over time. Also, there is an understanding that risk events can still happen even with pristine risk programs. Thus, it is equally important to recognize the roles of preparedness, recovery, and overall resilience, which we saw across many of the cases in this book.

While this book has related the importance of risk science from the lens of many case studies, unresolved issues also need to be addressed. Some of those issues can be addressed as the field grows, while others may never be addressed. These issues will be discussed in the next chapter.

16 Who Is the Risk Analyst and What Are the Expectations

This chapter will discuss the role of that risk analyst and potential repercussions to that individual or group of individuals. This chapter's discussion is based on Thekdi and Aven's (2023) work.

Across the many examples in this book, the logical questions to ask include: *Who is the risk analyst?* and *Who should be using the principles from this book?*

A natural answer to both questions is *everyone*.

The risk analyst can operate across a variety of professional domains. Consider engineers working to reduce risk associated with highway transportation or water quality. Similarly, engineers may work in manufacturing settings to ensure the quality of products or to understand the potential of failure for complex systems. Physicians diagnose and treat risk associated with patient health. Financial professionals assess and manage risk related to investment portfolios and in navigating regulatory compliance. At the policy level, consider the role of government officials in developing policies and contributing to decision-making for a wide variety of applications involving health, safety, and commerce. Teachers are risk managers as they promote mindful, informed, and productive citizenship for growing generations. In addition, law enforcement officers, emergency responders, social workers, and many others promote and engage in risk assessment, mitigation, and response, including during instances of natural disasters and other types of large-scale risk events.

While this list of risk analyst professional domains is not exhaustive, it is an example of the wide areas in which risk is understood and managed. However, one does not need a particular title or professional certification to assume risk analyst responsibilities. Everyone assumes those responsibilities in some contexts. For example, consider the groups described as follows:

Whistleblowers: Whistleblowers have a critical role in understanding and managing risk. Many past risk events had a reduced or minimized impact due to whistleblowers and their willingness to expose critical risk issues. The term *whistleblower* was made mainstream following the 2001 Enron scandal, after Sherron Watkins raised concerns about accounting irregularities, leading to increased scrutiny of the firm's practices. The impact of the whistleblower's decision was significant, as the event led to major regulatory reforms, including the Sarbanes-Oxley Act for corporate governance

DOI: 10.1201/9781003437031-16

and financial reporting. While whistleblowers are critical for addressing risk issues, there remains an opportunity for additional protection of those whistleblowers as they may face retaliation or social repercussions.

Informed citizens: The consequences of many risk events discussed in this book could have been much worse without informed citizens. For example, citizens who were quick to access information, share information, and respond to warnings during tornado events had a major role in risk mitigation. Informed citizens were also critical during the COVID-19 pandemic, acting in ways that potentially reduced the consequences of the virus. In a community setting, informed citizens aid in community watch programs, report potential risk issues in public infrastructure, and contribute to community discussions and decision-making related to public safety.

Promoters of health, safety, and associated practices: Individuals can promote health and safety in the most practical settings. They may do so using the appropriate public health guidance, societal safety norms, and knowledge of relevant laws and regulations. Some commonplace applications include checking smoke detectors and carbon monoxide detectors, reporting suspicious activities to law enforcement, following traffic safety rules, properly preparing for emergencies, and many other activities.

Despite the widespread role of the risk analyst, care should also be taken to understand the repercussions of the risk analyst's work. Even with pristine, high-quality risk analysis and decision-making that adhere to basic risk principles, risk events can and do happen. The risk analyst can sometimes face repercussions.

Consider the case of a 6.3 magnitude earthquake in L'Aquila, Italy, in 2009, which claimed the lives of over 300 individuals. Before the earthquake, the region had regularly experienced minor tremors. However, individuals responsible for risk evaluation and decision-making failed to identify these minor tremors as potential precursors to the impending larger earthquake. Subsequently, legal repercussions ensued for those individuals. Similarly, the Flint water crisis case study led to legal repercussions for those in risk analysis and management roles.

In these real-life cases, separating the risk analysts from the decision-makers is difficult. However, there is a realization that individuals responsible for risk, particularly those with professional responsibilities, can face repercussions. However, it remains true that the occurrence of a risk event does not prove or even suggest that a risk analysis, risk management, or decision-making regime was done improperly. Nonetheless, there are avenues for reputational failures, scapegoating, lawsuits, and professional ramifications.

Professionals across various industries often pledge to uphold integrity and adhere to their academic training principles. For instance, medical professionals take the Hippocratic oath, committing to treating the ill, respecting patient privacy, and consulting with specialists when necessary. Similarly, engineers pledge to support the health and safety of the public. Across professional settings, individuals and groups seek internal or external training and certification programs. Some commonly used programs include ISO 31000 and Enterprise Risk Management frameworks.

While there is a growing interest in risk principles in education and practice, there is no specific licensure or certification specifically for risk analysts.

Professionals engaged in risk analysis roles may encounter a range of repercussions, encompassing legal complexities and potential reputational implications. Legal issues may arise, leading to lawsuits or fines, particularly if individuals in fields like engineering or healthcare are deemed negligent in their duties. While insurance can provide financial buffers, those buffers do not undo real and potentially tragic consequences related to health and safety. Certain legal principles may offer protection in specific instances, such as in good Samaritan protections for healthcare workers in specific circumstances. As witnessed in events like the Flint water crisis or the L'Aquila earthquake, instances of legal trouble underscore the potential legal risks individuals face in risk roles. Companies, too, are not immune, and rules violations, as exemplified by the Sarbanes-Oxley Act, can result in significant legal consequences, including imprisonment and financial penalties. Addressing these issues, whether through legal proceedings or settlements, demands substantial resources in terms of time and finances. Even in cases of resolution outside the courtroom, associated costs, time, and stress persist.

Here, we describe specific factors that may have a role in the repercussions of the risk analyst.

Licensure: Risk analysts may need to adhere to specific protocols to secure and uphold professional licenses relevant to their job responsibilities. Examples of such licenses include medical licenses or professional engineering licenses. While specific licensing standards differ across professions and licensors, there are common reasons for the loss of license. For example, a common reason is malpractice or negligence. Other reasons could include misdiagnosis of a patient or situation, errors in performance of job responsibilities, failure to follow standard protocols, inadequate communication/documentation, or failure to maintain equipment. Supervisors may also face licensure issues in cases of failure to report particular actions that compromise safety or any other type of misconduct. Many of these features can exist within the realms of a risk analyst, particularly after a risk event has occurred.

Job duties/requirements: Risk analysts may be legally mandated to fulfill specific duties inherent to their profession. For example, consider the Upper Big Branch mine disaster. The industry was subject to the Mine Safety and Health Administration Regulations, while the risk event revealed many violations, including inadequate ventilation practices. Similarly, the BP oil spill noted legal mandates related to the Clean Water Act. At the individual level, workers in the nuclear industry are often required to document their activities, as imposed by regulators extensively. For example, workers are required to record details of training records, equipment inspections, maintenance activities, startup and shutdown procedures, equipment performance, and any anomalies or incidents. Failure to follow or accurately complete those requirements can be harmful for the risk analyst, through civil and criminal penalties, license suspension, and other serious consequences for the individual.

Public interest: Some risk issues and applications take on enormous public interest. For example, consider the issue of oil and gas drilling as related to the BP oil spill. There are many angles with which the public may engage with the risk issue. Concerns surrounding drilling relate heavily to environmental, ecological, and social issues. These drilling activities contribute to habitat destruction, oil spills, pollution, and climate change. Concerns over these issues are easily entrenched in the political sphere, promoting discussions and debates over energy security, job creation, and economic growth. In addition, discussions are raised about environmental conservation, climate action, and other issues. As described in Chapter 14, even the concept of tornadoes was politicized for its relationship with Cold War political tensions. In these types of situations, a well-meaning risk analyst, intending to perform their professional or otherwise neutral risk-related obligations, may inadvertently enter a political or social debate. That political or social debate may create enhanced scrutiny.

Publicity and associated motives: Some risk topics can be interesting for readers and may enhance interest in particular journalistic or other types of media outlets. We remind ourselves of the main features of an interesting journal article. A news article can be interesting for viewers when it addresses current and relevant topics with a human touch, involving personal experiences or stories. Conflict, controversy, and unique perspectives capture readers' attention, along with the visual appeal of images and videos. Because risk is about values that may differ across individuals and because risk involves uncertainties, it is not unexpected for risk issues to incite disagreement, conflict, or controversy.

In some cases, that public and journalistic interest can shed light on important risk issues that may not be known to the general public. For example, consider the Dhaka garment factory fire. For many, the publicity related to this risk event called for consumers to question the social and environmental footprint of their products.

In other cases, that interest can divert attention away from other high-priority risk issues, potentially in ways that promote misinformation and disinformation. A classic example of this is the event called the "Y2K bug." Leading up to the year 2000, there was concern and even panic about the potential impact of the Y2K bug on computer systems. This bug related to the way dates were represented in computer systems, particularly in systems using a two-digit year format. As the year 2000 approached, there was fear that computers might misinterpret the year as 1900, leading to errors and malfunctions. This concern was heightened by the widespread reliance on computer technology in various sectors like finance and utilities. The unknown consequences, potential economic and social impacts, and the need for precautionary measures attracted wide attention from the media and the general public. Significant efforts were made to address the Y2K bug and the fears of catastrophic failures did not materialize to the extent initially feared. While the attention toward the issue helped minimize the impact of the risk event, it has also been argued that the media sensationalized the issue by predicting catastrophic failures and disruptions.

Knowledge of the system: While this book emphasizes that there are vast uncertainties in forecasting future risk issues, there are also uncertainties related to understanding any system in a risk study. Knowledge related to these systems may also not be strong. However, communicating the limitations of weaknesses in knowledge can also be problematic because it may lessen perceived credibility. The Flint water crisis exemplifies how decision-makers struggled to define appropriate solutions due to insufficient knowledge of water quality issues within the studied system.

An emerging issue is understanding and developing knowledge related to systems reliant on artificial intelligence technologies. Because these technologies are poorly understood, even by engineers and scientists, there is concern over their use for risk applications. This remains an area of study for risk science and a variety of other domains.

Ethical issues: Risk studies are based on knowledge that comes from data, information, modeling, and argumentation. However, the decision-making component of a risk study is based on values, such as those related to health, safety, and the environment. Consider, for example, a common issue among many of the case studies of this book: the balance between profit and safety. For example, the BP oil spill and the Upper Big Branch mine disaster illustrate the balance between implementing safety policies and implementing those policies in a value-generating setting. Similarly, the case studies involving food additives show that some additives can prolong the shelf life of foods, despite demonstrating potential health hazards. This tradeoff between food accessibility and the potential for long-term consequences of those additives introduces ethical dimensions and debate.

Wellness implications: There may also be cases in which a risk analyst either has or perceives a sense of responsibility for the outcomes of risk decisions. Misalignment between decision-making outcomes and the analyst's values can weigh heavily on the risk analyst. In addition, the conditions in which the risk analyst performs their duties may contribute to burnout, lack of satisfaction in their role, and other wellness-related implications.

For example, consider the Space Shuttle Challenger disaster. Before the disaster on January 28, 1986, engineers expressed concerns about launching in unusually cold temperatures. The concern focused on the rubber O-rings used in the shuttle's solid rocket boosters, whose performance could be compromised by cold temperatures. Despite the engineers' recommendations to delay the launch due to safety worries, NASA managers, facing pressure to maintain the launch schedule, proceeded with the launch. Tragically, the O-rings failed during the launch, leading to the disintegration of the Challenger and the loss of seven crew members. While the engineers communicated the risk, the decision-making did not adequately heed those concerns. This marks an incident that could have been avoided. The repercussions of this disaster likely weigh heavily on all those involved. A news article (Berkes, 2016) detailing the memory of one engineer, Bob Ebeling, who attempted to publicize the critical risk and stop the launch stated:

> The morning of the launch, a distraught Ebeling drove to Thiokol's remote Utah complex with his daughter.
> "He said, 'The Challenger's going to blow up. Everyone's going to die,'" Serna recalls. "And he was beating his fist on the dashboard. He was frantic."
> Serna, Ebeling and Boisjoly sat together in a crowded conference room as live video of the launch appeared on a large projection screen. When Challenger exploded, Serna says, "I could feel [Ebeling] trembling. And then he wept – loudly. And then Roger started crying."

The factors discussed here also highlight the limitations of the risk analyst. While the analyst has control of the risk process, they do not necessarily have control over the outcome. The analyst may be a decision-maker. In other cases, the risk analyst cannot prescribe decisions for the decision-makers but does have some control over how information is presented. In some cases, even the decision-makers may not have influence over the eventual actions; such was the case of the COVID-19 pandemic in which decisions took place at many levels (e.g., policy, human behavior, and science) and those decisions were not in coordination with one another.

The risk analyst also cannot be seen in isolation from others, such as the public and other decision-makers. Thus, the work of the risk analyst includes both listening and communicating. Communicating aspects of uncertainty is challenging, as there is a delicate balance between acknowledging values and uncertainty involved with risk decisions while also seeking audience credibility and trust. The communication of risk issues can be nuanced in ways that involve many factors. Consider, for example, the Joplin tornado case in which sirens used for risk communication created confusion for some instead of spreading an important risk message of caution.

Finally, there remain many additional questions and debates surrounding ethical issues and how those ethical questions translate back to the role of the risk analyst. Elements of procedural justice, describing fairness in processes and policies regardless of the outcome, must also be seen alongside distributive justice, describing fairness in how costs and benefits are distributed. Within the domain of ethical principles is the task of understanding risk issues from a holistic perspective. Many cases in this book, including the Flint water crisis, involved many metrics of concern (e.g., cost-cutting initiatives, health/safety of residents, and politics). These metrics are not always factored into decision-making in impactful ways. In addition, many cases demonstrated a narrow focus on a limited number of stakeholders, thereby exacerbating the risk event in ways that impact a much wider variety of individuals. Thus, the challenge is for the risk analyst to manage their work's scope and content to consider these aspects, which is a challenging task.

Chapter 17 will further discuss the challenges in risk science by discussing unresolved issues.

WORKS CITED

Berkes. (2016). Challenger engineer who warned of shuttle disaster dies. https://www.npr.org/sections/thetwo-way/2016/03/21/470870426/challenger-engineer-who-warned-of-shuttle-disaster-dies

FURTHER READING

Thekdi, S. and Aven, T. (2023). Risk analysis under attack: How risk science can address the legal, social, and reputational liabilities faced by risk analysts. *Risk Analysis*, 43, 1212–1221.

17 Unresolved Issues in Risk Science Identified in the Presented Themes

While risk science provides a lens on many historical risk events, there are also areas in which risk science is still developing or intentionally silent.

First, there is the issue of ethical concerns. Across all of these case studies, there were ethical quandaries involved. For example, is it ethical to censor discussion about important risk topics even if those discussions are uncomfortable or induce emotional reactions? Is it ethical to stifle or promote narratives that are interpreted as misinformation or disinformation? Is it ethical for a risk analyst to promote particular risk decisions, knowing that the role of the risk analyst is to be neutral?

There is also little guidance on the role and the need for authenticity in risk regimes. Many existing regimes help organizations standardize their risk processes. Some large organizations discussed in this book had existing risk programs and processes in place before the risk event. However, the concern is that organizations may use their risk programs in cursory ways and not leave room for being sufficiently authentic to their own organization. Authenticity calls for organizations to carefully consider what the appropriate risk strategy is for themselves while meeting the related regulatory and certification requirements. For example, there may be instances where the organization chooses to hold higher or more detailed standards for risk-related protocols versus those required by regulatory standards.

Also, risk science has yet to clearly describe how one should measure the quality of a risk program when evaluating its past performance. While we can characterize the quality of a risk study, a risk program is much broader, including aspects of risk culture, interpersonal factors, data collection, and structure. There remains a vagueness in how to evaluate these types of aspects. In addition, we recognize that the adoption of risk science principles does not guarantee the absence of a risk event. Risk events, including surprises, can happen, whether the risk is mitigated, accepted, transferred, or avoided. Conversely, when risk events do not happen, it's not clear whether that absence of risk events is due to appropriate risk management, luck, lack of reporting, or some other factor. Thus, adequately understood and managed risk may appear hidden or quiet.

The next big frontier of risk comes from the ever-increasing use of modern technologies. The concept of explainability refers to the ability of new technologies, such as those associated with artificial intelligence and machine learning, to provide understandable and transparent explanations for their decisions, actions, and outputs. Initiatives for explainability are in response to growing concern about the lack of transparency in these emerging tools, as they are often referred to as "black boxes,"

DOI: 10.1201/9781003437031-17

as their inner workings are not easily understandable to the end-users or even the developers themselves. Initiatives for explainability can include elements such as describing how various models arrived at their decisions, identifying the explanations for why models failed or were erroneous, uncovering potential biases in decision-making, and other emerging forms. While these issues generally have existed for some time, serious risk and ethical concerns remain.

Another large fronter is understanding how to consider issues like equality, equity, and fairness within a risk context. For example, consider the case of the Texas power grid failure in 2021. The impacts of the power outages and lack of essential services fell disproportionately on some populations. The impacts of the failure severely impacted populations living in housing that were more susceptible to the impacts of extreme weather conditions, lacked resources to address power outages, and freezing temperatures, and may have been vulnerable from a health perspective. As another example, consider the cases of H1N1 and COVID-19, in which the risk management measures impacted younger populations and their access to schooling. There remains a lack of guidance on how these issues are factored into risk-based decisions. There is also a question of the role of risk science with respect to these issues as those issues relate heavily to political and policymaking dimensions.

Another area of inquiry for risk science is distinguishing noise from relevant information that should guide risk-based decisions. Broadly, across the case studies of this book, it was recognized that there was significant hindsight bias. After risk events occur, one can ask what signals were missed or misinterpreted. Preceding the risk event and even during risk events, it is unclear what information is relevant and actionable. Often, there are attempts to provide counterfactual discussions, suggesting that if there was some differing reaction or decision, the risk event could have been avoided or had a much less severe impact.

Consider an example of hindsight bias from the perspective of a major assassination in history.

ABRAHAM LINCOLN ASSASSINATION – HINDSIGHT BIAS VERSUS NOISE LEADING TO A RISK EVENT

Abraham Lincoln, the 16th President of the United States, was assassinated on April 14, 1865, by John Wilkes Booth. The assassination took place five days after Confederate General Robert E. Lee surrendered at the Appomattox Court House in Virginia, marking the end of the American Civil War. While attending a performance at Ford's Theatre in Washington, DC, Booth, a confederate sympathizer, shot Lincoln. The president was taken to a nearby house, where he died the following morning. Booth, who sought to avenge the defeat of the Confederacy and perceived injustices, managed to escape the scene initially but was eventually tracked down and killed after a massive manhunt. Lincoln's assassination had a profound impact on the nation and marked a tragic moment in American history.

Lincoln faced considerable political opposition. History has recorded many death threats, assassination plots, and attempts at sabotage throughout his political career. Lincoln employed widely varying mechanisms for safety, including secrecy of travel plans, sudden changes in itinerary, attempts to block public view, traveling in disguises, disrupting telegraph communication to deter information flow about his travels, limited public appearances, and using bodyguards. For example, consider the Baltimore Plot, an alleged conspiracy to assassinate President-elect Abraham Lincoln as he traveled through Baltimore on his way to Washington, DC, for his inauguration in 1861. It was rumored that secessionist groups from Maryland and other southern states planned to kill Lincoln during his stopover in Baltimore. In response to the threat, Allan Pinkerton, a private detective and Lincoln's security chief, devised a secretive plan, which included changing travel plans to avoid the danger in Baltimore. The plot heightened tensions and highlighted the seriousness of the secessionist sentiment, prompting increased security measures for the President-elect, particularly leading up to the Civil War (Figure 17.1).

These measures to address the serious risk of assassination show apparent attempts toward risk characterization and risk management activities. In hindsight, one can speculate about the risk characterization and management deficiencies that contributed to the later assassination. The event was a surprise to many for several reasons. For example, in terms of evaluating security at Ford's theater, it was unexpected that an actor would pose a threat. At the time, the president's bodyguard was off-duty and was substituted by a different guard. Theories suggest that the president offered the guard a better seat, leaving himself less protected.

Hindsight bias can promote questions like: How could the assassination been thwarted? Would the bodyguard have been able to prevent the attack? Why did the security team not consider security at the theater amidst the barrage of threats? Why did Lincoln choose to go to the theater despite an earlier premonition? Did Lincoln choose to accept the risk instead and reduce the emphasis on risk management

FIGURE 17.1 A photo of Lincoln at the Gettysburg Address. One can see closeness of crowds. Wikimedia Commons (1863).

activities? As he famously said: "If I am killed, I can die but once; but to live in constant dread of it, is to die over and over again."

While the answers to those questions cannot be known, many have attempted to answer them. Yet, at the time, these factors were largely seen as noise. The importance of these factors, such as threats from actors and standards for security, are now apparent, only by looking to the risk history for guidance.

This chapter has shown that risk science is constantly improving and still has areas for further development in research and practice However, looking across a broad range of historic events allows us to identify the major areas in which risk science can evolve. Only through the study of history can we explore initiatives on innovations that will prompt improved understanding and attention toward a broad range of new and emerging risk issues.

The following chapter will provide general conclusions, including how to learn more about risk science and where to go from here.

WORKS CITED

Wikimedia Commons (1863). File: Lincolnatgettysburg.jpg. https://commons.wikimedia.org/wiki/File:Lincolnatgettysburg.jpg

18 Conclusions

The authors of this book often ask students: "What is the difference between an accident and an incident?" Students often look to the dictionary to identify the differences between those two terms. We can ask whether events that appear to be accidents, in retrospect, are incidents that resulted from inadequate or insufficient assessment, management, and communication of risk. Across this book, we have encountered a mixed set of cases that includes accidents and incidents with large potential for debate when making those distinctions. Risk science allows us to use a more structured approach to making those distinctions.

While this book is a collection of case studies, there are several additional resources that can be used to learn more about risk science. The authors have published the book *Risk Science: An Introduction*, which serves as a textbook for graduate risk programs across the world. The authors have also published *Think Risk: A Practical Guide to Actively Managing Risk*, which discusses nontechnical risk science principles for a broader audience.

This book has discussed many best practices related to risk science. There is an opportunity for readers to discuss these principles with groups of peers in professional and personal settings and ask how risk decisions are currently being made versus what improvements can be made. Are there any biases that are detrimental to initiatives to address risk? Are there areas in which risk processes can be improved? In addition, in a workplace setting, there is an opportunity to investigate the existing risk programs and learn more about those existing practices. The dissemination of information about those programs should ideally be openly discussed, further creating a more comprehensive and effective risk culture.

There is also an opportunity for readers to carefully consider the risk-related implications of shared and received information. For example, readers can consider whether the information they share or consume on social media constitutes misinformation or disinformation. Readers may also call out when misinformation or disinformation exists, when appropriate. Additionally, readers may carefully consider the integrity of information they encounter in popular media, asking questions about the quality and transparency of evidence used in various risk-related claims. The expectation of high-quality risk studies further creates opportunities to better understand and manage risk in the future.

Finally, the authors hope that this book has helped understand risk science as it relates to historical events. While risk science does not offer answers to all questions, it provides a framework to understand risk in ways that can be generalizable across industries and professionals. It is that ability to branch across the many aspects of risk that will allow future abilities to address risk holistically.

DOI: 10.1201/9781003437031-18

FURTHER READING

Aven, T. and Thekdi, S. (2024). *Risk Science: An Introduction*. London: Routledge.
Society for Risk Analysis. (2024). https://www.sra.org/

Index

Pages in *italics* refer to figures and pages in **bold** refer to tables.

A

acquired immunodeficiency syndrome (AIDS), 8
AI algorithms, 92
aleatory uncertainty, 39
American Civil War, 113–114
anchoring bias, 74
Ashe, Arthur, 8
Atomic Energy Commission, 41
Atoms for Peace speech, 40–41
availability bias, 73

B

Baltimore Plot, 114
biases in risk science literature
    anchoring bias, 74
    availability bias, 73
    confirmation bias, 74–75
    and moral hazards, 76–79
    optimism bias, 75
    publication bias, 75
    representativeness bias, 73–74
    single action bias, 75–76
Boeing 737 Max crises, 10
BP oil spill, 108, 110
    causal factors related to well blowout, 85
    Deepwater Horizon incident on May 22, 2010,
        83, *84*, 85
    drilling activities, 108–109
    ESG initiatives, 86
    regulation, 85–86
    repercussions, 85
    risk assessments for oil drilling
        deepwater drilling, 84
        skimming operations, 83, *84*
        uncertainties, 84
British White Star Line, 33
Bush, George W., 15, *15*, 51, *52*

C

California climate crisis, 14
censorship, 64
classical probability, 38
communication
    clear, 61, 69
    of risk, *see* risk communication
    of severe weather, 63
    uncertainties, 62–63

competing narratives, 80
confirmation bias, 74–75
consequences
    forward-looking projections for, 33
    as prediction of future, 31
    in risk assessment process, 33
    short- and long-term, 37
        Flint water crisis projection, 36
        Titanic case evaluation, 33–37; *see also*
            Titanic case study
    VSL/willingness to pay, 32
COVID-19 pandemics, 7–8, 14, 21–22, 28, 70
    global, 7–8
    individual behaviors, 9
    misinformation and disinformation, 79–80
credibility of knowledge
    food additives case, 58–60
    new/emerging evidence, *see* evidence
Cuyahoga River fire, 65–71
cyberattacks, 54
cyberbullying, 101

D

data
    and associated modeling, 39
    collection, 3, 11, 45, 51, 56, 94, 112
    and information, ix, 3, *4*, 5, 7, 23, 28, 45, 48,
        57–59, 67
deepwater horizon oil disaster, 9–10, 83, *84*, 85;
        see also BP oil spill
Dhaka garment factory fire, 10–11, 24, 31
dioxin poisoning, 11
disinformation, 5, 28, 73, 77–80, 95, 100–101,
        103–104, 109, 112, 116

E

East Palestine train derailment, 12–13, 73, 76–79,
        104
Ebeling, Bob, 110
Eisenhower, D., 40
Emergency Response Coordination Center
        (ERCC), 91
Enterprise Risk Management (ERM), 21, 64, 75,
        107
environmental events, 83
environmental impacts, 89
Environmental, Social, and Governance (ESG), 86

EPA's seven cardinal rules, 61–65
epistemic uncertainty, 39
equality, 9, 22, 58, 71, 82, 113
equity, 9, 21–22, 58, 66, 71, 82, 113
ERCC, *see* Emergency Response Coordination
     Center
ERM, *see* Enterprise Risk Management
ESG, *see* Environmental, Social, and Governance
evidence, 78
     about food additive, 59
     analytical perspective, 57
     in aspects of high-quality risk study
          analysis, 58
          data and information, 57–58
          managerial review and judgment,
               decisions and communications, 58
          overall risk study, 57
     from data, 57
     groups interpret, 56
     high-quality, 56
     high-quality evidence, 56
     individuals interpret, 56
explainability, 112–113

F

Factory Acts legislation in the United Kingdom,
     19
Failure Modes and Effects Analysis (FMEA),
     39, 43
falsehoods, 22, 25, 28, 48, 73, 77–80, 95,
     103–104
FBI, 50–51
February 1993 World Trade Center bombings, 50
Finley, J.P., 94–95
Flint sit-down strike of 1936–1937, 19–20, *20*
Flint water crisis, 108–109, 111
     case study, 47–49, 107
     infrastructure failures, 11–12
     short-and long-term consequences, 36
food additives case, 16
     packaging, 58–60
     regulations over use of titanium dioxide, 59–60
food-grade titanium dioxide, 59–60
Ford's theater, 113–114
Foreign Intelligence Surveillance Act, 53
frequentist probability, 38
Fukushima Daiichi Nuclear Accident, 13–14, 28
     case, 43–45
future events, viii, 1, 24–25, 39, 103

G

Garrick, John, 1
general knowledge (GK), 47–49, *49*
General Motors (GM) automobile factory, 19
Gulf Coast, 89, 91

H

high-quality risk studies, 56–58, 69, 86, 103–104,
     107, 116
hindsight bias, 24, 113–115
historical events
     acts of terrorism and war
          main features, 15
          September 11, 2001 (9/11), 14–15;
               see also 9/11 terrorist attacks
     categories, 7
     food safety
          food additives, 16
          forever chemicals, 16
          main features among cases, 16–17
     global pandemics, 7
          COVID-19, 7–8
          HIV/AIDS, 8
          main features among cases, 8–9
     industrial accidents
          Boeing 737 Max crises, 10
          Deepwater horizon oil disaster, 9–10
          Dhaka garment factory fire, 10–11
          main features among cases, 11
          Seveso disaster, 11
          upper big branch mine disaster, 10
     infrastructure failures
          East Palestine train derailment, 12
          Flint water crisis, 11–12
          main features among cases, 12–13
     natural disasters
          California climate crisis, 14
          Fukushima Daiichi Nuclear Accident, 13
          main features among cases, 14
          Texas energy grid failure, 13–14
HIV/AIDS, 8
hot boxes, 76–77
Hurricane Isaac, 91
Hurricane Katrina, New Orleans
     before *vs.* after damage, 89, *91*
     flooding in New Orleans, 89, *90*
     resilience efforts, 90
     response to Hurricane Maria, 92
     response to Isaac, 91
     rooftops/shelters, *89*, 89–90
     storm, 89
Hurricane Maria, 92

I

Industrial Revolution, 19, 98
information; *see also* disinformation;
          misinformation
     characteristics of, 78
     dissemination of, 116
     risk-related, 65–71
insurance and investment analyses, 83

integrity, ix–x, 3, *4*, 27, 64, 78, 80, 103, 107, 116
International Conference for Safety of Life at
        Sea, 35
Ismay, J Bruce, *35*
ISO 31000 framework, 107

J

Joplin tornado case, 94, 96, 111

K

Kaplan, Stanley, 1
knowledge, 29
    classification of, 46, *47*, 48
        9/11 case, 49–54; *see also* 9/11 terrorist
            attacks
        judgments about strength, 39, 44–45
        justified beliefs, 46, 56
        knowledge/understanding, specific group A
            and general B
        known known (Known to A, known to
            B), 48
        known unknown (known to A, unknown
            to B), 47
        unknown known (unknown to A, known
            to B), 48
        unknown unknown (unknown to A,
            unknown to B), 48
    new, 28
    risk characterization and decision-making, 103
    suppression/manipulation of, 101
knowledge/beliefs, 26–27, 29
known unknowns, 46–47

L

Laherrère, 28
Lake Huron source, 11–12
L'Aquila earthquake, 107–108
Lee, Robert E., 113
Lewis, H.W., 41–42
Lincoln, Abraham, 113–115, *114*

M

Macondo well/blowout, 84–85
malinformation, 77–78
Maneuvering Characteristics Augmentation
        System (MCAS), 10
media, 5, 12, 23, 47–48, 57, 62, 65, 69–70, 96,
        109, 116
    attention, 71
    forms of, 101
    social, *see* social media
Merchant Shipping Act of 1894, 34
Mercury, Freddie, 8

metaphors, 27–29, **29**
metrics
    environment, 32
    objective function, 31
        profit term in, 32
    single standard, 33
        dollars, 32
    Societal Health, 32
Mine Safety and Health Administration
        Regulation, 108
misinformation, 73, 77–80
modern scientific knowledge, 103
moral hazards, 76–77

N

NASA, 39, 43, 110
The National Commission on the BP Deepwater
        Horizon Oil Spill, 83
National Incident Management System (NIMS), 91
National Institute of Standards and Technology
        (NIST), 97
National Research Council (2014), 43
National Weather Service (NWS), 95
natural disasters, viii–ix, 7, 14, 36, 70, 83, 88, 92,
        98, 106
    California climate crisis, 14
    Fukushima Daiichi Nuclear Accident, 13
    main features among cases, 14
    Texas energy grid failure, 13–14
New Orleans, 89–92; *see also* Hurricane Katrina,
        New Orleans
9/11 Commission Report (2004), 50–51
9/11 terrorist attacks, 14–15, 21, 26–27, 29
    Bush, 51, *52*
    February 1993 World Trade Center bombings,
        50
    issues of national security, 49–50
    knowledge between groups, 49–50
    Obama, Barack, 53, *53*
    USA PATRIOT Act, 51
    War on Terror, 51
nuclear power, 13–14, 39–41, 43, 45
nuclear testing, 95–96, 103

O

Obama, Barack, 53, *53*
Offshore Drilling report (2011), 83
Oil Pollution Act of 1990, 85
optimism bias, 75

P

past risk events, x
    Fukushima Daiichi nuclear disaster case study,
        74

interpretations, 23
   quality control initiatives, 24
   remember and understand, 23
per- and poly-fluoroalkyl substances (PFAS), 16,
   70, 73–74
personal protective equipment (PPE), 76
Phoenix Memo of July 2001, 50
Pinkerton, Allan, 114
popular epidemiology, 48, 54
PRA, *see* Probabilistic Risk Assessment
The President's Daily Brief (PDB), 50
probabilistic risk assessment (PRA), 20–21
   at nuclear plants
      disadvantages, 44
      knowledge judgments, 44–45
      results of, 44
      using levels, 43
   probability, 39–40, *40*, 43
   of nuclear incident, 44
publication bias, 75
public health crisis, 12–13
Puerto Rico, 92

R

radar technology, 94
railroad safety, 13, 73
Reagan, Ronald, 8
reality, 56, 58, 62, 65, 71, 82, 103
Red Dye No. 3, 16
Renn, O., 69
representativeness bias, 73–74
resilience
   concept of, 88
   Hurricane Katrina, 89–92
   importance of, 92
   modern technologies, 92
   for securing critical infrastructures, 88
risk
   (A, C, U), 1
   authenticity in risk regimes, 112
   concept, 2, 6
   contexts, 1, 24, 113
   culture, 21, 43, 104–105, 112, 116
   evaluation of, 22, 107
   evolution of, 22
   exposure, 17, 70
   factories
      acts, 19
      Flint sit-down strike of 1936–1937, 19–20,
      20
   methodological, 21
   PRA, 20–21
   principles, 18, 86, 107–108
   quality of risk program, 112
   quantification, 20, 22
   safety concerns of Industrial Revolution, 19

societies, 18
structured frameworks of ERM, 21
understanding and managing, viii, x, 1, 19–21,
   36, 64, 76, 86, 105–106
   future risk, 24, 29
   Titanic case, *see* Titanic case study
   unfamiliar issues, 71
   various aspects, 18
risk analyst
   professionals, 107–108
   responsibilities
      informed citizens, 107
      promoters of health, safety and associated
         practices, 107
      whistleblowers, 106–107
   role in repercussions
      ethical issues, 110
      job duties/requirements, 108
      knowledge of the system, 109–110
      licensure, 108
      public interest, 108–109
      publicity and associated motives, 109
      wellness implications, 110
risk applications, 82
   concept of VSL, 82
   insurance industry, 82
      at corporate level, 83
      at operational level, 83
   risk dimensions for oil and gas operations and
      regulation, *see* BP oil spill
risk assessment, 27
   characterization
      (A′, C′, P, SoK, K), 1–2, 31
      (A′, C′, Q, K), 1, 31
      A types of events and A′, 1–2, *2*
      consequences, *see* consequences
   process, 31, 33
   questions of, 1
Risk Assessment Review Group, 41
risk communication, viii, x, 5, 9, 16
   Cuyahoga River fires in Cleveland, Ohio,
      USA, 65–69, *66–68*
   EPA's seven cardinal rules
      accept and involve the public as a
         legitimate partner, 62–63
      be honest, frank, and open, 64
      coordinate and collaborate with other
         credible sources, 64–65
      listen to the public's specific concerns,
         63–64
      meet the needs of the media, 65
      plan carefully and evaluate your efforts, 63
      speak clearly and with compassion, 65
   personal, 68–69
risk perception
   controllability, 70
   dread, 70

familiarity, 71
media attention, 71
tornadoes, *see* tornado
trust, 70
voluntariness, 70
SARF, 69–70
risk events, 1
absence of, 112
characterizations of uncertainties, x
and consequences, 2–3
credible knowledge, x
criticism, 43
future likelihood, 24–25
history/historical, 104
categories of, ix–x
documentation, viii
product of interpretation, ix
integrity of knowledge, x
Y2K bug, 109
risk management, viii, x, 113
decisions, 9
individual behaviors during pandemics, 9
process
consequences, 5
data and information, 5
decision-making and communication, 5
knowledge, 5
uncertainty, 5
risk-related biases, 73; *see also* biases in risk
science literature
risk related information, 65–71
risk-related messages, 65, 69
risk science, 80, 112
concepts of, x–xi, 103
explainability
concept of, 112
initiatives for, 112–113
high-quality, 25, 86
history for, ix
practical application of, xi
principles and practices, viii–ix, 22
process for general analysis/science, *4*
professional analyses in, 3
risk analyst's concern, 104
risk characterization, 4
risk management decisions, 5
understanding and managing risk, viii–x, 1,
19–21, 36, 64, 76, 86, 105–106
unresolved issues in, x
Risk Science: An Introduction, 116
RMS Carpathia, 34, 36
Rumsfeld, Donald, 46–47, 50

S

safety culture, 9, 11, 77, 86
safety equipment, 75

Safety of Life at Sea Convention agreement in
1914, 36
safety-oriented technologies, 76–77, 86
Sarbanes-Oxley Act, 106–108
SARF, *see* Social Amplification of Risk
Framework
scientific research, 45, 59, 94, 98
Senate Investigating Committee, *35*
September 11, 2001 (9/11), *see* 9/11 terrorist
attacks
Seven Cardinal Rules of Risk Communication,
61–65; *see also* risk communication
severe acute respiratory syndrome coronavirus 2
(SARS-CoV-2), 7
Seveso disaster, 11, 73
Shippingport Nuclear Power Station, 40
Signal Corps, 95
single action bias, 75–76
skepticism, 20, 28, 45, 56, 69
smoking, 98–102, *99–100*
Social Amplification of Risk Framework (SARF),
69–70
social media, 98–102, 116
glamorization of, 98, *99*
nature of, 101
platforms, 101
Societal Health metric, 31
societies, viii, 1, 18, 21, 23, 98
Sornette, 28
southern United States, 94
Space Shuttle Challenger disaster, 110
specific knowledge (SK), 47–49, *49*
stakeholder engagement, 21
Stokes, Carl, 66, *67*
storms, 13, 27–29, **29**, 57, 61, 89–90, 92, 94–96,
98, 102
strength of the knowledge (SoK), 31
subjective probabilities, 38–39
surprise and unpredictability
black swan concept, 26–27, **29**
dragon-king, 28, **29**
one's knowledge/beliefs, 26–27
risk events
9/11 terrorist attacks, 26–27, 29
perfect storm, 27–29, **29**
system resilience, 3, *4*, 88

T

Taleb, N. N., 26
terrorism, x, 7, 92
cyberattacks, 54
and war, 50–51
September 11, 2001 (9/11), 14–15, *15*;
see also 9/11 terrorist attacks
terrorist attacks, viii, x, 7, 21, 26–27, **29**, 46, 50–51,
92; *see also* 9/11 terrorist attacks

Texas energy grid failure, 13–14, 113
Think Risk: A Practical Guide to Actively Managing Risk, 116
Titanic case study, 75
    attention on grandeur and luxury, 34
    British White Star Line, 33
    head of the White Star Line, 35, *35*
    passengers in lifeboats, 33–34, *34–35*
    RMS Carpathia, 34, 36
    scenarios and consequences during ship design, 36
titanium dioxide as food additive, 58–60
tobacco, 98–99, *99–100*
tornado(es)
    concept of, 102
    1955 radar imagery of, 61, *62*
    practices in tornado-prone areas, 96
    prediction, 95
    preparedness, 96
    related to risk attenuation and risk amplification, 69
    risk perspective, 61
    risk science in action, 94
    tilt-up design, 97
    in weather forecasts, 61
    weather warning system, 95
transparency, 32, 41–42, 58, 65, 78–80, 102, 112, 116
Tri-State Tornado of 1925, 94
tsunami, 13, 28, 43, 70

U

uncertainty characterization, 31
    aleatory uncertainty, 39
    epistemic uncertainty, 39
    failure of nuclear power stations likelihood of early fatality, 41, *42*
    Fukushima Daiichi case, PRA at nuclear plants, 43–45; *see also* probabilistic risk assessment
    1979 Three Mile Island accident, 43

probability
    classical, 38
    frequentist, 38
    PRA, 39–40, *40*, 43
    in risk assessment, use of, 39
    subjective (knowledge-based, judgmental), 38–39
    of system failures, 39
    WASH-1400 nuclear reactor safety study, 40–42, *42*
United Nations General Assembly, 40–41
United States Department of Homeland Security, 15, 21, 53–54
United States Environmental Protection Agency (EPA), 61
The United States National Security Agency, 51
United States Navy SEALS, 53
unknown unknowns, 46, 48
upper big branch mine disaster, 10, 24, 108, 110
USA PATRIOT Act, 51
U.S. Center for Disease Control (CDC), 8
U.S. Clean Water Act, 85
U.S. Nuclear Regulatory Commission (NRC), 40–41

V

value of a statistical life (VSL), 32, 82
VSL, *see* value of a statistical life

W

Wagner Act of 1935, 19
War on Terror, 51
WASH-1400 nuclear reactor safety study, 40–42, *42*
Watkins, Sherron, 106
weapons of mass destruction (WMD), 51
Weather Bureau in the United States, 61
World Health Organization, 8, 101
World Trade Center in New York City, USA, 14–15, *15*, 51
World War II, 41, 98

Printed in the United States
by Baker & Taylor Publisher Services